SHAFT FURNACES AND BEEHIVE KILNS

A History of Smelting in the Far West
1863-1900

SHAFT FURNACES AND BEEHIVE KILNS

A History of Smelting in the Far West 1863-1900

By Lynn R. Bailey

Lynn R. Bailey

Westernlore Press . . . 2002 . . . Tucson, Arizona

Library of Congress Control Number 2001134027
ISBN 0-87026-114-2

Edition limited to
1,000 copies

PRINTED IN THE UNITED STATES OF AMERICA BY WESTERNLORE PRESS

CONTENTS

ILLUSTRATIONS

ILLUSTRATIONS (Continued)

FOREWORD

In the quest for precious metal between 1849 and 1870 large deposits of base metal were discovered throughout the American West. The first were copper deposits in the western foothills of the Sierra Nevada and at Williams' Fork on the Colorado River. Those strikes were followed by discovery of silver-bearing lead at Cerro Gordo in what would become Inyo County. About the same time, immense bodies of silver-lead were located at White Pine and Eureka in Nevada. More deposits of lead were found in Utah; in the mountains surrounding Salt Lake City and southward in Beaver County. Because these lodes carried varying amounts of gold and silver, they could not be treated by amalgamation, and miners were left no other alternative but to work the ore by smelting.

In the 1860s, when the first base metal lodes were exploited, smelting was a little understood art, and there were few metallurgists in the Far West. Nevertheless, miners attempted to reduce their ore, building Catalonian furnaces and vasos, both of great antiquity. Others utilized blast furnaces used in iron foundry work. While these primitive furnaces were cheap to construct, and in most cases worked, their production was limited, and smelting faltered.

Other men, however, stepped forward to make improvements. European metallurgists and ore buyers, drawn to California's mineral wealth, applied Welsh and French methods to copper deposits in Calaveras, Plumas, and Fresno counties. And in 1863 a resourceful foundry operator named Nathaniel Haskell modified an iron-smelting cupola by putting a water-jacket around the crucible. Application of a water-jacket to a furnace, the first in North America, had far-reaching consequences. That innovation was carried back to Europe and applied to the German smelting industry at Freiberg and Clausthal. Metallurgists returned that idea to the American Far West a few years later. Hired to open silver-lead lodes in Nevada, German smeltermen such as Joseph Mosheimer, Guido Kustel, Otto Hahn, Charles Stetefeldt, and Albert Arents built large masonry, high capacity water-jacketed blast furnaces to process Nevada's silver-lead lodes. These German-designed Piltz and Raschette smelters were soon in use in Utah, Montana, Colorado, and Idaho, spawning America's smelting industry.

As advanced as these furnaces were, they were expensive to build and run, and could not be moved when ore deposits vanished. That deficiency was corrected by a Welshman, John R. Williams, Jr., who with his father and brothers, took the best German ideas and incorporated them into a portable wrought-iron furnace, which could be moved from one location to another. Permitting the exploitation of small, rich carbonate deposits, the Williams or Rankin-Brayton Furnace, as it was called, revolutionized lead and copper smelting in Nevada, Utah, and Arizona, and every base metal locality worldwide.

In short, base metal smelting in America's Far West is the story of innovation, experimentation, and modification. From a humble beginning, smelting opened a vast store house of wealth, created population growth, and gave birth to other industries such as the charcoal trade, freighting lines, railroads, sampling works, foundries, and firebrick manufacturers. In developing America's resources, the smelting industry turned small towns into metallurgical centers where reduction techniques and equipment were tested and perfected.

In putting this story together, I have been aided by several people. Don Chaput, curator emeritus of the Los Angeles Natural History Museum, provided material on Copperopolis and Cerro Gordo, as well as biographical information and data pertaining to equipment manufacturers. And Tucson journalist Dexter Oliver contributed photos of charcoal kilns located near Milford, Utah.

LYNN R. BAILEY
Tucson, Arizona
September 17, 2001

SHAFT FURNACES AND BEEHIVE KILNS

A History of Smelting in the Far West 1863-1900

A charcoal blast furnace, ca. 1852. All furnaces used to reduce base metal ores were modifications of the ordinary iron blast furnace, shown above. Ore, charcoal, and fluxes (if required), were fed to the furnace in alternate layers, at a distance of 10 to 12 feet above the level where jets of air were forced into the furnace. As the charge of charcoal and ore moved downward, it became heated, first to redness, and subsequently to the melting point. The heavy metallic alloy (lead, copper, gold and silver) sank to the bottom of the furnace. The lighter slag, floating upon the molten metal, was drained off through a slag spout, and discarded. The metal, called bullion, was drawn off at intervals from the bottom of the furnace.

Chapter One

COPPER SMELTING IN CALIFORNIA

Reduction by fire to obtain metals from ore is an ancient and widely practiced technology. But it was not introduced to the Western United States on a commercial scale until 1864, fifteen years after the discovery of gold in California. Reasons for this delay are numerous, the most obvious being that native gold, which existed in great quantity, was easily retrieved by placer mining. The quest for free gold lasted several years, and led to discovery of lodes and veins carrying both gold and silver. Rock was broken, crushed, and chemically treated to extract the precious metals; the quartz mill came into vogue and great technological strides followed. The chemistry of amalgamation, known for 400 years, was perfected and proclaimed as the principal ore reduction method in the Far West. Discovery of large deposits of base metal — lead and copper carrying varying amounts of gold and silver — created a new era in metallurgy on the Pacific Coast. To extract silver and gold from base metal called for smelting.

Smelting was a little understood practice, although feeble attempts had been made to smelt iron at two locations, the earliest being in Utah. In winter of 1849-50 "a hill of richest iron ore" was discovered by a group of Mormons led by Parley P. Pratt in the region of the Little Muddy, in what would become Iron County. In response, Brigham Young organized the "Iron Mission," and in late 1850 and early 1851 two parties of Mormons, many of them converts from iron-manufacturing centers of England, Scotland, and Wales, pushed south from Salt Lake City to establish Parowan and Cedar City, and exploit Utah's iron resources.

In late 1852 they built a blast furnace, assembled large quantities of ore and charcoal and poured a small amount of pig iron — the first pig iron produced west of the Mississippi River. The enterprise was reorganized as the Deseret Iron Company and capitalized at $40,000. An improved blast furnace was built, and in March 1853, 2,500 pounds of "clear good iron" was made. Indian troubles and a destructive flood destroyed the enterprise later that year. The colony was rebuilt in 1854, and iron production was under way in earnest the following year, the iron going into the manufacture of nails, agricultural implements, household wares, and machinery parts.

Just when it looked like the Iron Mission would achieve success, nature dealt a blow. Winter of 1855-56 closed canyons and shut off the supply of charcoal, and freezing weather curtailed the flow of water. Costly furnace blow-outs the following year interrupted production, and drought and a grasshopper infestation finished the enterprise. It was not revived until after completion of the transcontinental railroad in 1869. By then others were producing pig iron in the Pacific Northwest.

In 1866 W. S. Ladd of Portland, Oregon, in partnership with New York and San Francisco businessmen, incorporated the Oregon Iron Company to smelt brown hematite mined at Oswego, on the left bank of the Willamette River, eight miles south of Portland. Under direction of George D. Wilbur, a veteran iron master from Sharon, Connecticut, the company built a massive basaltic rock blast furnace, 32 feet high and nine feet in diameter, capable of smelting 3,500 tons of pig iron per annum. Because coal was unavailable, kilns were erected and charcoal produced from wood harvested on company property. On August 27, 1867, the Oswego furnace poured its first pig iron, the first produced on the Pacific Coast. While the quality of iron was excellent, the enterprise failed because of high labor and fuel costs.[1]

The next stab at smelting on the Pacific Coast was with copper. Although copper smelting, like the reduction of iron in Oregon, got off to a wobbly start, it laid the foundation for subsequent base metal smelting in the American West. It is a little known, yet fascinating story that began early in California history.

Copper was known to exist in California as early as 1840, having been discovered in Soledad Pass, 90 miles north of Los Angeles. Traces of the red metal were found north of San Francisco Bay in 1848, and J. B. Trask, who acted as state geologist from 1851 to 1854, found copper in nearly every county, his first observation being made in Nevada County at a place called Round Tent.[2]

In 1855 a prospector named J. W. Bean picked up specimens of copper carbonate from a ravine near his cabin in Hope Valley in Calaveras County, on the western slope of the Sierra Nevada, not far from the Placerville and Carson Valley road. When his partner remarked that the specimens "were nothing but copperas rocks, the country is full of the stuff, and it isn't worth a d—n," Bean threw the samples out the door with the rubbish. In 1860 Hiram Hughes, a former resident of Calaveras County, while prospecting in Nevada recognized formations at Washoe as being similar to those in the California foothills. He returned to Calaveras County and located claims on Quail Hill, a prominence on a slate belt about 35 miles above Stockton, and six miles from the Stanislaus River, opposite the Central ferry.

Assaying from $4-$7 per ton gold and $5-$20 silver, the vein was taken up for precious metal. Nuggets of native silver were initially found in ravines crossing the vein, but it was sulphuret of copper, with a show of carbonate on the surface, that wetted the appetites of prospectors. It did not take long for the vein to be proclaimed the "Comstock" of the copper region.[3]

News of the Quail Hill discovery quickly spread, drawing a horde of prospectors hunting for "indications." They were followed by San Francisco capitalists. Among the latter was W. K. Reed, who examined Hughes' claim in summer of 1860. Satisfied that the copper ore held promise, Reed prospected eight miles to the east of Hughes' claims. On Hog Hill in the Gopher District, he discovered copper mineralization. He gathered specimens and retraced his steps four miles to McCarty's cabin, a well-known stage station on the Stockton Road, where he submitted them to a Dr. Blatchly, of San Francisco, who had recently been attracted to the region by Hughes' discovery.[4]

An amateur mineralogist, Blatchly pronounced the specimens copper "of most flattering indications." The next morning Reed, McCarty and Blatchly located and staked off for themselves and friends 11,250 feet of mineral-bearing land that would later become the Union, Keystone, Consolidated, Empire, Napoleon, and Calaveras mines. A large number of sacks were filled with carbonate ore gathered from the surface. The first work was done by Blatchly on the Keystone claim. In what was called the Discovery shaft he struck carbonate ore at 25 feet. In the meantime, Reed commenced work upon the Union claim, obtaining good ore at about the same depth as found by the Doctor. Running about 30 percent copper, the ore was shipped to San Francisco, loaded aboard ocean going vessels, and conveyed to Swansea, Wales.[5]

Satisfied there was permanency to his copper discoveries, Reed in fall of 1860 erected a canvas house under a big oak tree, which served as a boardinghouse for employees of the Keystone Mine. A small wooden building a few feet from Reed's tent was the next house built. Reed also erected a boardinghouse for Union Mine hands. Other wood and canvas structures followed. Early in spring 1861 J. M. Bean erected the Copperopolis Hotel, one of the largest structures in the region. Building of that edifice created quite an excitement, and a general staking out of lots followed, and the town of Copperopolis was born. From that time on the town grew rapidly. Within 18 months Copperopolis boasted three hotels, several restaurants, two livery stables, two barber shops, six to eight well filled stores, and numerous establishments where "creature comforts" were dealt out. A school house was erected and a lot of ground set aside for a cemetery. By summer 1863 Copperopolis presented all the peculiar characteristics of early mining camps of 1850, with the exception that its houses were constructed of lumber rather than canvas, as in the early days. The camp's population was estimated at 1,200, with Cornish and Irish miners forming "quite an item."[6]

The Copperopolis boom was not a localized incident; it was tied to an interest in copper that had gradually heightened over the decades. In 1835 the aggregate production of copper worldwide was only 25,000 tons annually; the price in chief markets being $350 per ton. From that time until 1855 the yield gradually rose to 55,000 tons per year, and the price from $450 to $475 per ton. By 1865 the price of copper stood at $500 and the annual yield amounted to 70,000 tons, or $34,000,000 a year as the price of worldwide copper production. Obviously demand exceeded supply.

Before the Electric Age copper was combined with zinc to make brass, which entered into every metallic art, from the manufacture of a watch or kettle to a cannon. Bell metal was made of copper and tin, as was bronze for statues. Tons of copper was used in currency in both Europe and America. Copper also formed part of the composition of knife handles, harness ornaments, and went into the service of every household or trade or art from the highest to the lowest. Merchant vessels and warships were "copper bottomed."

Up to 1863 Great Britain and Chile furnished more than half the world's copper, about 14,000 tons annually. The United States produced 2,000 tons, but Cuba, Austria, Australia and Russia exceeded the United States in copper production. In 1853 copper deposits in the Lake Superior region began to attract attention and capital,[7] but for the most part American copper mines were scattered and poorly run.

With demand outstripping the supply of copper, news of sizeable deposits of the red metal in California created a mining boom, causing the value of copper claims to skyrocket. Less than a year after the copper strike in Calaveras County, W. K. Reed sold six shares, or 975 feet of the Union Mine, for $65,000 to Meador, Lolor & Company of San Francisco. Subsequent development of the mine inflated the value to $1,200 per foot or $1,140,000 for the interest sold by Reed. Capitalists from New York, Boston, and Swansea, Wales, opened offices in San Francisco to purchase ore.

By January 1863 there was considerable traffic in Calaveras County ore. Shipping merchants were advancing 50 percent on 15 percent ore, basing the price on specie value in New York or Boston. The currency value being $5.75 to $6, to the one percent of the assay which, on 15 percent ore amounted to between $86 and $90 per ton. Ore of 12 percent and higher in copper content, mostly from the Napoleon Copper Mining Company, was being delivered at the "ship's tackle" in San Francisco, payable in gold coin. Shipments of copper ore from San Francisco to New York and Boston between January 1 and August 17, 1863, amounted to 4,600 tons, valued at $500,000. By August a dozen mines were regularly shipping ore to San Francisco. Only the richest carbonate ore, assaying above 12 percent, paid to be shipped. Ore below that tenor had to be "worked," which presented problems.[8]

In 1863 there were only three ways of working copper ores: smelting in reverberatory furnaces (the Welsh process, a laborious method involving 12 to 15 steps); smelting in blast furnaces, and treating by lixiviation or leaching. In France, Germany, Sweden, Norway, and Russia blast furnaces were generally used. In England, Spain and Chile reverberatory furnaces were utilized; and in the eastern United States both smelters were used, although blast furnaces were preferred. Reason for the different smelters being that work was performed according to conditions of price, quality of fuel and labor. In England where good fuel was cheap, and experienced labor abundant but paid little, the reverberatory furnace flourished.

In California wages were high, labor was scarce, and experienced metallurgists were practically non-existent. Thus, the early years of copper reduction in California was a period of trial and error.

The first plans to erect smelting works to reduce California copper ore were laid by Thomas Price, a 28-year-old Welshman who had come to San Francisco in 1862 to purchase ore for the Vivian Company of Swansea, Wales. A trained metallurgist, Price saw opportunity in the California copper boom beyond that of just purchasing ore. He felt lower grade copper ore could be reduced by smelting to regulus, a matte of about 60 percent copper, and then shipped to England or the Eastern United States for further refining. He intended to do this by application of the Welsh process with which he was familiar.[9]

With the view of erecting reduction works, Price and a number of wealthy San Francisco capitalists incorporated the California Smelting Works in spring of 1863. The company purchased a tract of land at Antioch, on the south bank of the San Joaquin River, two miles above its juncture with the Sacramento River. The location was judicious, lying within five miles of the Mount Diablo coal mines, and directly on the route over which passed all ore from the region drained by the Sacramento and San Joaquin rivers.[10]

Under supervision of Thomas Price, the company commenced buying ore from mines in Calaveras, Nevada, and Yuba counties at the rate of about $2 for each percent of copper the ore contained. For 11 percent ore they paid $22; and $25 for 12 percent ore. Accumulating 2,500 to 3,000 tons of ore, Price and his associates put up a reverberatory furnace. Built to produce a matte of 50 percent copper, this reverberatory of eight ton capacity was 13 feet, six inches long by six feet, four inches wide, with a 65-foot stack. Coal from Mt. Diablo fueled the furnace.[11]

The company, however, did not intend to carry refining through the labor intensive twelve or more steps of the Welsh process. Instead it would convert the ore into "regulus," an operation performed as follows: "A given quantity of the different ores are turned into the furnace, the average of which must not assay over 10 or 12 per cent. The whole is well heated, and in the course of a few hours the furnace is tapped and the entire mass drawn off, in the condition of what is termed regulus, which will assay from 40 to 60 per cent of copper."[12]

The California Copper Smelting Works was running by fall

1863, but had to be shut down every 15 days to reline the furnace. About September 14 the company shipped a bar of regulus, "the first copper ever obtained outside of a crucible on the Pacific Coast." Pouring of this copper bar, asserted the *Mining and Scientific Press,* also proved "the suitableness of Mount Diablo coal for smelting copper," an assertion not entirely true.

At any rate, on September 21, 1863, the *Press* proclaimed a new industry had been introduced to the State. "The entire California public are watching with much interest the developments which this company is making in a branch of industry, which if successful, will open up an almost unlimited field for the employment of capital, and add millions of dollars to the value of a most important class of mines."

So satisfied was management of the California Copper Smelting Works that another large furnace was erected in October 1863. In the meantime, the California press touted their success, proclaimed a bright future for the State's brass industry. The company even envisioned supplying copper to China and "various other places."[13]

The first furnace runs, however, did not meet expectations; the regulus assayed no more than 30 percent copper. To raise the tenor, a calciner was added, and the ore roasted to get rid of sulphur. The first charges from calcined ore, mixed with copper carbonates and oxides, produced a regulus assaying 75 percent copper; too rich. Several other charges produced 60 per cent. The average regulus assayed 46 percent; still too low. In spring 1864 the smelter shut down while the trustees squabbled over chemistry. The problem was solved and the smelter was in full blast by April 1865, producing regulus of forty percent copper. A second fusion reduced the regulus to black copper (85 to 90 percent). In the end, the California Smelting Works failed, not because of its inability to smelt copper, but because of the high cost of financing its operation. At that time interest on money in California stood at 18 to 24 percent.[14]

Nevertheless, interest in California copper continued to grow, heightened by two events. The first was the 1863 discovery of lenses of copper carbonate and red and black copper oxide in schistose and basaltic rock at Sailor's Diggings in Josephine County, Oregon, just over the California border. Much of this ore, in veins 15 to 25 feet wide, was extremely rich, running between 30 and 60 percent copper. To exploit the Oregon discovery, a French metallurgist and professor of chemistry, Dr. Paul D'Heirry, organized the Queen of

Bronze Copper Mining Company. Incorporated in April 1865, and capitalized at $792,000, the company sunk a shaft and drove a 238 foot long tunnel. D'Heirry erected two cupola furnaces of a French pattern, and proceeded to refine ore according to the German method, which was based on the use of blast furnaces.[15]

Mining and smelting copper in Oregon may not sound like much, but when D'Heirry's published a series of articles in the *Mining and Scientific Press* he stirred up considerable interest in copper production. Entitled the "Metallurgy of Copper in California," he outlined techniques employed by the Queen of Bronze Copper Mining Company, and detailed furnace construction.[16] Advocating the use of native stone instead of firebrick, which had to be imported from England or Baltimore, the Frenchman made smelter building sound simple, within reach of miners provided they were directed by metallurgists.

A conflict between Spain and Peru was the second event spurring interest in copper. When the Spanish fleet blockaded Peruvian ports in fall of 1865 the price of red metal shot up by $150 per ton. Copper in Liverpool on October 16, 1865, was $396.88 a ton; one week later, on October 23, a ton was being quoted at $416.24; and on November 3 the metal was selling for $464.64. In anticipation of further advance, New York and Eastern manufacturers bought heavily, causing the *Mining and Scientific Press* to predict that "California will . . . reap a permanent benefit from these troubles."[17]

The paper was right, California's copper output soared. During 1864, 14,000 tons of copper ore from Calaveras County, valued at about $600,000, passed through Stockton. That amount increased in

D'Heirry's furnace was similar to the one pictured at right. Called a *Fourneau a manche,* it was used throughout France. Its base was made of solid masonry, strengthened by transverse bars of iron. The cavity was lined with a coating of refractory material, cemented to the masonry in the form of a ellipse, and renewed every season. The two lateral faces were constructed of gneiss, and the front, called *fiervende,* of rectangular iron plates, coated with fire clay. The rectangular cavity was six feet long, five and a quarter broad, and three and a quarter deep. The tuyere had an orifice of three inches in diameter, its nozzle being wrought and its bed cast iron. Opposite was a platform made of clay firmly beaten, in which there was dug a crucible on a level with the sole of the furnace. That was coated with a mixture of clay and finely powdered charcoal, and from it passed an inclined canal leading to the receiver in which molten metal collected.

Above: Blast furnace elevation: *b b*, exit openings; A. shaft. Below: the sole of the furnace: BB, basins, *b* tap-door, *c c* channel to conduct metals to furnace, tt tuyeres. (From Piggot)

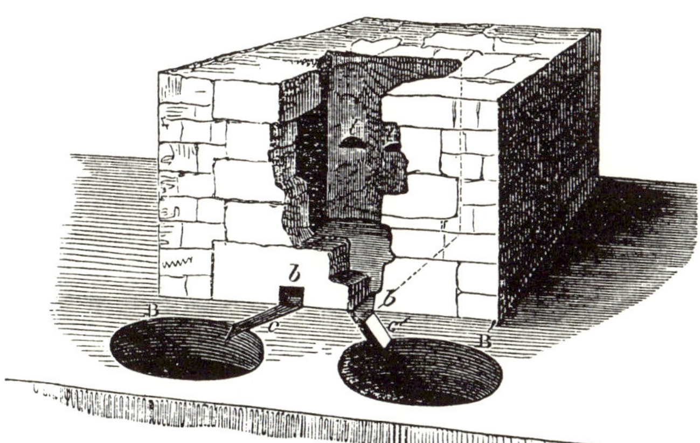

1865 to 25,000 tons valued at $1,500,000. During this same period R. B. Parker & Company shipped 2,828,881 pounds of copper ore to San Francisco from the Keystone Mine at Copperopolis. These shipments marked the beginning of Calaveras County's copper business.[18]

The Keystone Mine at Copperopolis, under the direction of the Powning Brothers, would figure large in California's copper trade. The Powning Brothers, five in number, were Cornishmen. Miners all, as was their father and grandfather before them. One brother was at the Buri-Buri mines in Australia, another worked in the copper mines of the Lake Superior region, and still another brother had charge of the Carey mine at Washoe. Two other Pownings were in Calaveras County, superintending the Keystone, Calaveras, Napoleon, and Josephine mines.

The Pownings felt that Calaveras County in particular and California in general would benefit from the importation of "miners who were bred to the business and understood it in all its details; men who could not only mine and dig, but men who could sort and select ores after they were mined." The Powning Brothers induced about 100 Cousin Jacks to come to Calaveras County. They were the first Cornishmen to apply their brand of mining technology to the region.[19]

In early 1864 the Pownings erected concentration works at the Keystone Mine, the first of its kind on the Pacific Coast. The *Mining and Scientific Press* believed them to be "pioneer works of this Continent." Designed and built by Henry Tregellas of Donahue, Booth & Company, of San Francisco, the concentrator was housed in a building 200 feet long by 35 feet wide. The 65-horsepower engine had a 15-inch cylinder and 48-inch stroke. The six-foot diameter boiler was 25 feet long, with return side flues, which provided about 600 feet of heating surface.

The ore for concentrating was carried from the shaft to mill by means of a railway and hoisted to the crusher room over an incline, where it fell into a hopper and passed between two heavy chilled cast rollers. The crushed ore then passed through a revolving screen; the finest material sifting into sluice boxes, the coarser particles being returned by means of a rait wheel to be crushed again. The finest particles were carried into the concentration house and passed through five jiggers composed of a tank about five feet long by two and a half feet wide, nearly filled with water. Inside these tanks were sieves attached to a cam-shaft above. The ore was shoveled into the sieves, and by a quick up and down motion, slate and other waste matter was lifted to the surface, scraped off and thrown away, the concentrate put into bags for shipping to smelters then being built in San Francisco.[20]

In 1864 there was considerable metallurgical experimentation going on in San Francisco. That year C. E. Johnson patented a smelting process, and in partnership with a Mr. McDougal, erected a smelter at North Point near San Francisco, spending about $10,000 on the project. Another project was being carried on by a German metallurgist named Joseph Mosheimer, who had engaged in ore reduction in the early days of the Washoe discoveries, and studied ore reduction methods on the Continent and in Great Britain between 1861 and 1864. Upon his return to San Francisco he opened a metallurgical office on Bryant Street and advertised his ability to plan and erect furnaces and reduction works. About that time, another Frenchman entered the smelting act.[21]

In September 1864 Pierre Desormeaux took over the superintendency of the copper smelting works of Meader & Company, a San Francisco and Stockton mercantile firm which owned the Union Mine at Copperopolis. He erected two cupola furnaces similar in design to D'Heirry's in Oregon. Smelting with charcoal without addition of flux, Desormeaux in October produced a matte, "sometimes richer and sometimes poorer," but overall "very satisfactory." By April 1, 1865, he was pouring 4,000 pounds of copper regulus per day. According to the *Mining and Scientific Press* Desormeaux was the first person to profitable work "refuse copper ores."[22]

By September 21, 1865, the Union Mine had smelted and shipped 807 tons of regulus. The *Stockton Independent* noted the shipments of ore and regulus, elaborating that "the copper trade of Stockton is of such importance that in addition to what is carried as freight by the regular steamers, the Navigation Company now keep the steamer *Helen Hensley* and the barges *Sacramento* and *J. Bragdon* exclusively employed in the transportation of copper ore."[23]

The next region to attract the attention of skilled smelters was Campo Seco north of Copperopolis, up the Salt Spring Valley in Calaveras County. Four distinct copper-bearing veins, over four miles apart, existed there. By end of 1863 hundreds of claims had been staked, the most important of which was the Campo Seco, located at Copper Hill on the banks of the Mokelumne River. Opened to a depth of 200 feet, this claim had a vein of sulphides of iron and copper, 23 feet wide, one-sixth of which assayed from 12 to 23 per cent copper. In 1865 the Campo Seco Copper Mining Company, a joint stock company, was formed. The firm paid expenses by shipping nearly two million pounds of copper carbonate to San Francisco for processing in local metallurgical establishments, or turned over to agents for refinement at Swansea, Wales. Lower grade ore was worked at smelting works close to the Campo Seco Mine.[24]

Those works were located on the Mokelumne River, three-quarters of a mile below the mine. Christained the Taunton Smelting Company, this reduction plant was patterned after the Crocker Brothers' works at Taunton, Massachusetts, and the Point Shirley Works of Boston, the only American smelters of copper ores at the time using upright or stack blast furnaces. De Heirry's furnace in Oregon was of similar design.

Writing in the *Engineering and Mining Journal* of December 4, 1886, W. H. Adams described the 1865-66 Crocker Brothers works at Taunton, Massachusetts, which processed ore mined at Capelton, Canada:

> These furnaces were modeled from those in use in England, Germany, Spain, Norway, etc., . . . built of red brick, usually 6 feet by 8 feet and from 10 feet to 12 feet high, topped out with a low chimney, and held together by iron stay-bars. The furnace proper was fire-brick, 2 feet by 4 feet inside, with breast of tiles held in place by cross-bars and springs. The bottom of the furnace was 3 feet from the floor level, and blast was supplied through one tuyere opening placed 8 inches above the bottom at the back of the furnace. Outside the furnace front, crushed coke and fire-clay ("steep") were banked up compactly, and basins were scooped therein, into which the molten slag ran for separation from any overflowing matte. After completion, these furnaces, costing perhaps $200, were slowly dried out, and the week's campaign started by firing with coke Sunday night. Monday morning, at seven, blast was put on and ore charging regularly followed until Saturday at four p.m., when the furnace was completely cleaned out for renewal. The charges of fuel and ore (1:4) were carried up in small boxes and thrown into the furnaces over the breast tiles, the burden above the tuyeres being about 5 tons.
>
> The slag, after leaving the "steep," flowed into sand-beds, and was removed in slabs when cooled. The matte or regulus

gradually formed a basin inside the furnace, by eating out the sides and bottom brick, and then oveflowed into the "steep" basins, being tapped from them into sand-beds or ladled into molds. The product therefore was more or less impure by mixture with slag and sand. At least once a day, the entire "steep" front was removed, the matte was drawn from the interior of the furnace, a new front made, and blast continued. The campaign of the week was satisfactory if 40 tons of ore were smelted, or 40 ore, 10 slag, 10 fuel equal 60 tons put through in $5^1/_2$ days. The matte was from 20 to 25 percent from low-grade ores, and the slags carried from $1^1/_2$ to 3 per cent copper.

The works were placed in a central position to all the company's property, to water, and fuel for steam purpose. . . . The smelting building was 120 feet by 40 feet and 12 feet clear between floors. A building in the rear contained the 70 horse-power boilers, engine, blowers, smith-shop, etc., and an addition to the east side was for 8 matte stalls.

The ore and fuel were thrown through the roof doors . . ., and [a] stock of 200 tons could be carried under cover on the level with the feeding-doors. Water was brought in wooden logs, bored on the place, from a reservoir 100 feet above. Slags were disposed of on a short run; matte was sent down to the stalls, and down to wagons for shipment.

The smelting building was completed with four stacks, three for ore and one for matte-smelting. The latter, however, was never used, as the others were equal to all demands.

Each furnace was expected to run 15 tons of roasted ore daily, and was built on red brick piers, 7 feet high, with outside measurements, 7 feet by 9 feet; height to feed-doors, 11 feet; chimney raised above the roof, and all work securely bound by iron stay-bars. Inside the piers, the furnace proper was of fire-brick with expansion-joints in contact with red brick. The bottom of the furnace was 3 feet above the floor, and the breast was built of tiles 5 feet above the bottom, the front being closed above that point. Two tuyeres were placed in line at the back 8 inches above the bottom of the furnace.

Outside the furnace front, three cast-iron plates formed a box to hold the "steep" basins, and had lips for the overflow of the slag, and tap-holes for matte on each side. The matte was received in rectangular shallow cast-iron molds, free from dirt or slag, and easily cooled by elevation on rails. Slag was removed in ordinary cumbersome wagons.

Two of these furnaces ran from 12 to 18 tons of ore daily for $5^1/_2$ days, and were pulled out for repairs according to the prevailing fashion. Three months training of the men devel-

> oped the fact that the damage to the lining of the furnace took place the first three days, and it therefore was deemed possible to continue the campaign; but here came the difficulties of custom, training, stubbornness, etc., to overcome, and a struggle it was for weeks. A run of three weeks was made by bribes, threats, and withdrawing one furnace for rebuilding purposes, which demonstrate the principle and gave encouragement for renewal of one furnace with 3 tuyeres in the back.
>
> These tuyeres . . . cast in Sherbrooke, were 3 feet long, 15 inches across base, 2 inches across nose, and 2 inches water space. Although much trouble was occasioned by the burning out of . . . the tuyeres, causing explosions, sows, etc., yet this furnace ran six weeks with an average daily consumption of 23 tons of ore (fuel 1:8).
>
> Another furnace, changing to same pattern, ran eight weeks with every condition better, and part of this time ran at the rate of 35 tons of ore daily. These two furnaces were run for nearly a year with improvements in tuyeres, tools, tapping facilities constantly going on. . . .
>
> The work became so regular that No. 2 furnace was torn down and rebuilt for 5 tuyeres placed in the back . . . which furnace ran with all ease from 40 to 50 tons daily. With very dry, well-rosted ore, 60 tons have been put through in 24 hours.[25]

For its time the Taunton Smelting Company at Campo Seco was a sophisticated plant, consisting of two furnaces, the blast created by a McKenzie blower. The McKenzie blower, patented in 1855-56 by New York machinist Phillip Wallace McKenzie, was a rotary pressure blower designed for use in iron smelting, the blast of which was created by rotation of a shaft carrying a number of blades. Having no valves, this blower was dependent upon steam engines or water wheels, the power transmitted by belt or gearing. In the case of the Taunton Smelting Company, the McKenzie blower was driven by a water wheel.

A horse-drawn railroad connected the mine and smelter. Brought to the smelter site, the ore was first roasted in immense heaps to drive off the sulphur. Six men, besides outside laborers, were employed in this process, which cost about one dollar per ton.

The furnaces were blown-in under supervision of a Mr. West in August 1865. According to the *Mining and Scientific Press* the smelter worked "like a charm," each furnace consuming daily from 200 to 250 bushels of charcoal, costing 20 cents per bushel. During one furnace run of 43 days in fall of 1865, 119 tons of matte was

A belt driven fan blower made by the B. F. Sturtevant Company of Boston, Massachusetts. The McKenzie blower was similar in design. (From Birkinbine, "American Blowing Machinery")

produced. Operations, however, were halted when charcoal could not be obtained in sufficient quantity to run through the winter.[26]

In early 1865 another smelter fired up in upper Plumas County, a "grand, magnificent" area of pine covered mountains, and vast granite ridges two to three thousand feet high, through which wound the grassy Genesee Valley. A copper rush boomed here in spring of 1864, with claims being staked on every indication of copper along Grizzly and Genesee creeks. San Francisco capitalists rushed in, buying shares in claims by the foot. A town was laid out near the mouth of Grizzly Creek, and a saw mill, boardinghouse and several other buildings were constructed. In summer of 1864 Laufman and Company built a smelter.[27]

The real power in Plumas County, however, would be the Genesee Valley Copper Smelting Works, located about five miles from Taylorville, built by Bolinger, Blood & Company, which also owned mines in the area. Designed and constructed by S. Platt, the $30,000 smelter was supervised by John C. Chapman, one of the mine owners said to be an "experienced copper smelter."

Built after the French plan described by A. Snowden Piggot in *The Chemistry and Metallurgy of Copper*,[28] this smelter consisted of a furnace house large enough to accommodate four furnaces, a coal house, blacksmith shop and boardinghouse. Two blast furnaces were initially built, the blast driven by two double acting piston bellows or tubes, four feet in diameter, the whole being actuated by a 25-horsepower overshot water wheel. This smelter was "constructed on the principle of running copper direct from the ore, without the use of any flux other than the slag and metal from previous meltings and a judicious mixing of different ores."

The furnaces were not large, fueled with charcoal made from pine wood, and charged with 750 pounds of ore, ranging from eight to 13 percent, with an average of 11 percent. Fired up in mid-November 1864, the first furnace run failed. After three hours and fifteen minutes under blast, it was found dampness had "raised in the hearth," causing the metal to adhere to it, raising the heat above the fire-proof lining, and melting the upper lining which was of common brick. The result was only 40 pounds of copper regulus. The furnace was repaired, charged again, and the temperature brought to operating point. After three hours the blast was applied. The furnace was tapped and a 15 pound pig of copper produced. Shortly thereafter the front furnace lining, constructed of common brick, again melted down. Repaired this time with a species of asbestos rock, the furnace was in operation by mid-December 1864.

The furnace ran successfully for some time, for the *Mining and Scientific Press* reported on November 18, 1865, that the smelter was turning out "on the average about 600 pounds of pure copper daily. . . . The melted copper is not drawn off into molds as at Copperopolis, but, as it runs from the furnace into a vat kept hot by a surface of burning coals, it accumulates till the vat is full. Then the coals are removed, and the worthless slag soon cools in a crust and is then thrown away; next successive crusts of matte, of about fifty per cent of copper are thrown off at the rate of three for each minute. This is kept for a second refining process. Next the pure copper appears at the bottom, and is known by the fact that no crust will form on the surface when the whole is left to cool in a mass of pure metal. The freight upon pure metal to San Francisco is comparatively light, where it is sold at a good profit to all engaged in the work."[29]

While the Genesee Valley excitement was going on there was

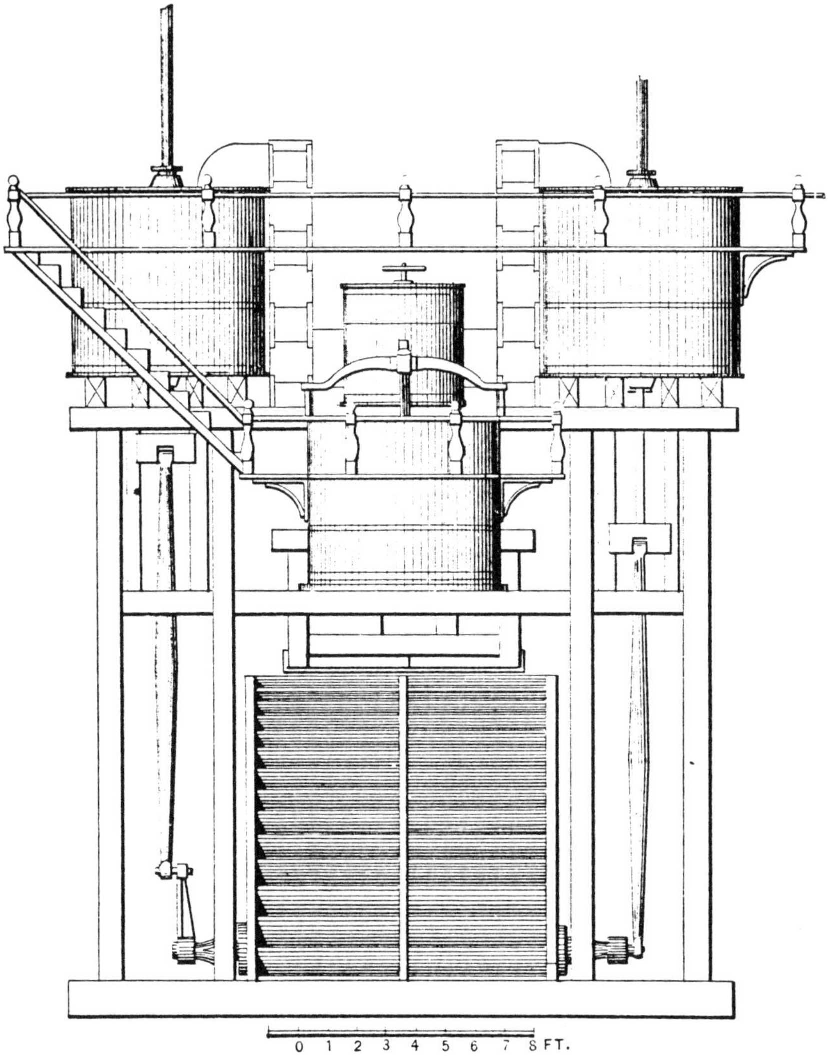

The blast of the Genesee Valley Copper Smelting Works was supplied by blowing tubes similar to those pictured above. Made of wood, braced or banded, these tubes, ordinarily operating in pairs, were from four to seven feet in diameter. The pistons had short strokes, seldom over two feet. They were driven directly from a water wheel shaft. The capacity of a pair of wooden blowing tubes seldom exceeded 1,000 cubic feet of air per minute, delivered under one-half to one pound pressure. Such blowing tubes were used in charcoal iron furnaces of the 1840s. (From Birkinbine, "American Blowing Machinery")

great activity in a 14 mile long, three to four mile wide, copper-bearing belt laying in Mariposa and Fresno counties. Here occurred a siliceous slate and sandstone formation rich on the surface with copper carbonates and silicates with a small admixture of red oxides, and in some cases large masses of gray copper. A few claims were opened, but not to any great depth. Two furnaces were erected in this district: the first being that of Nathaniel Haskell, who had a foundry at Mariposa. The second was Henry & Company's smelter.

Henry & Company's furnace was located on the Lone Tree vein at James' Ranch in the Hamilton District of Mariposa County, a copper bearing zone that was hoped would rival Copperopolis. By January "quite a little town" had sprung up in the locality. Patterned after the blast furnace described in Dr. D'Heirry's articles on "Copper Smelting in California," this smelter was erected in late 1865 by Pierre Desormaux, formerly of Meader & Company's smelting works at Copperopolis. The furnace, which reduced ore from the Lone Tree and Green Mountain leads, burned in October 1865 and was replaced in February 1866 with a "new and extensive smelting works," consisting of two furnaces. Operational in April, these smelters produced 180 bars of copper matte in a two day run. Another day and half run turned out 75 pigs of copper. Declared "a complete success," two more furnaces were erected which brought the smelting works' daily capability to 16 tons of silicate and carbonate ore averaging 12½ percent. The smelters, employing eleven men, consumed five to six tons of charcoal daily, costing $17.50 per ton, making the cost of smelting copper from $75 to $85 per ton. Unfortunately, the James' Ranch copper deposit had no depth, mining ceased, and the smelters were shut down.[30]

It was much the same story for a Mariposa foundry operator named Nathaniel Haskell who, during the copper excitement of 1865-66, procured a claim on the Buchanan lead,[31] in Buchanan Hollow at the head of Chowchilla Creek, six miles below James' Ranch, in the upper part of Fresno County. In partnership with a man named Phillips, Haskell designed and built a furnace to process copper not only from his claim but others in the district. His enterprise is significant in the history of West Coast smelting.

Although Haskell's furnace, also called the Buchanan Company, was a simple cupola such as used in foundries for melting iron, it was somewhat of a revolutionary device. The furnace crucible or cupola was cast at the Golden State Iron Works in San Francisco.

Risdon and Coffey, a Bay City boiler making shop, created a wrought iron water-jacket, which prevented breaking, cracking and warping from heat, so common in ordinary furnaces. Probably sectional in design, it was claimed the furnace could be put up in ten hours after its components had arrived on the ground. The "Haskell" iron cupola-furnace, of about four ton daily capacity, consisted of a sheet-iron shell provided with a thin fire-proof lining. According to Albert Arents, the designer "connected steam-tight the ends of two boiler-shells differing in diameter, left openings for blast-pipes at proper places, as is customary with furnaces for smelting pig-iron, put a drop-bottom underneath, filled his space between the boiler-shells with water . . . and smelted his copper ore." This was the first application of a water-jacket to a furnace in America.[32]

Haskell's furnace was fired up in January 1866, and after some experimentation, ran successfully for 21 days, with a net profit of one thousand dollars a day. Five thousand pounds of copper were shipped in February 1866, making a total of 24,000 pounds shipped since completion of the furnace. In a one week run in mid-January Haskell produced a little over five tons of copper. In another 60 day run, 17,612 pounds of copper, 90 percent pure, was produced. The *Mariposa Gazette* of April 7 reported "Thursday last, Mr. Haskell operated with his smelting works twenty-seven hours, and the product of copper, in pigs, during that time, was *two tons and a half*. We doubt whether similar results can be shown since the organization of the State." The copper pigs from Buchanan Hollow were carried by wagon 30 miles to the San Joaquin River and loaded aboard steamers and transported to San Francisco. To publicize the efficiency of Haskell's furnace, a pile of copper bars were kept on display at Coffey & Risdon's shop on Market Street.[33]

In the end, the water-jacket of Haskell's furnace sprung a leak, the smelter failed, and in June 1866 the tenor of ore changed from carbonate to sulphide, and then gave out completely. Shortly thereafter the smelter was transferred to the Crane & Brigham smelting works at Hunter's Point, near San Francisco. That was not the end of Haskell's furnace, however.

Manufactured by Coffey and Risdon, a number of Haskell furnaces saw brief service elsewhere in the West. One was transported to Nevada to smelt lead-silver ore, and several were shipped up the Colorado River to the Freeman District where pockets of copper carbonate had been discovered on the west side of the river opposite

Williams' Fork. In fall 1866 the Planet, Knowles and Lightner, and the Great Central companies, all in the Freeman District, employed Haskell furnaces to reduce ore that would not bear the expense of transportation. From all indications the furnaces functioned well. The Great Central Copper Mine, which fired up their Haskell furnace on April 6, 1867, reported that "in two hours after turning on the blast, the copper was running like water." It was shut down, inspected, and fired up again. The furnace proved a success, smelting from 16 to 20 tons of ore each day.[34]

About 1870 the price of copper fell by two-thirds halting all copper mining. Most Haskell furnaces were abandoned, others were moved elsewhere to be used in lead reduction. Out of work, Haskell moved south to Sinaloa, Mexico, to raise sugar cane. Haskell's furnace, however, did not fade away; it had features that could not be ignored.

Shortly after the original Haskell furnace was transferred to Hunter's Point (in 1866) a young German metallurgist named Albert Arents inspected it and wrote an article describing its water-jacket in the summer 1866 issue of *Berg-und Huttenmaunische Zeitung*, a German mining journal. A year later the water-jacket principal was employed on furnaces in the Harz region of Germany. Although the idea for a water-jacket was conceived as early as 1852, it being mentioned in relation to iron smelting on page 556 of Frederick Overman's *Treatise on Metallurgy* (Overman died in 1852, prior to publication of his work), Haskell may have unknowingly influenced smelting worldwide.[35]

* * *

1. A good discussion of early attempts to smelt iron in Utah is Gustive O. Larson's "Bulwark of the Kingdom: Utah's Iron and Steel Industry," *Utah Historical Quarterly*, Vol. 31, No. 3, Summer 1963, pp. 248-261. A more recent treatment is Janet Burton Seegmiller, *A History of Iron County*. Salt Lake City: Utah Historical Society, 1998, pp. 73-82.

The Oregon Iron Company was incorporated February 23, 1865, with a capitalization of $200,000. Blast for its mammoth furnace was furnished by blowers, no more than wooden cylinders and pistons, driven by water power. Although high cost of fuel and labor shut down the furnace for a time, it was revived, and twice enlarged. The furnace was in blast almost continuously for 10 years. In 1878 the plant was bought by the Oswego Iron Company, which removed the wooden appliances and operated the furnace intermittently until October 26, 1882, producing in four years 18,500 tons of pig iron. In 1882 the plant was taken over by the Oregon Iron and Steel Company, which produced 14,663 tons of pig iron in the two succeeding years. Some of the iron made during this period sold as high as $40 a ton. Between 1885 and 1889 the company was reorganized with a capital of $1,500,000 and a new furnace was built in 1888, the old furnace being abandoned. The new furnace was operated until 1910. *Mining and Scientific Press*, December 22, 1866; *Engineering and Mining Journal*, July 23, 1910, p. 173.

2. Hubert Howe Bancroft, *History of California*, Vol. VII, San Francisco: The History Company, 1890, pp. 658-659.

3. *Mining and Scientific Press*, August 24, 1863.

4. There is another version of the Copperopolis discovery. According to the *California Farmer* of August 21, 1863, "the copper mines were first discovered on the 29th of July, 1860, by W. K. Reed, who was then harvesting on the farm of T. Hardy. Copper has been found in the Gopher district, and had been located in the name of Mr. Hardy. During harvest word had been sent to Mr. Hardy to send a man over to Gopher to do work sufficient to hold the claim, and Mr. Hardy sent Mr. Reed. It was in passing to and fro from Hardy's ranch to the Gopher district, that Reed discovered the famous Union lead. Mr. Reed had been a miner in the lead mines of Wisconsin; he had been long a miner here, yet never a lucky one, but always the reverse, until he struck the Union lead. When he made the discovery he set to work in earnest with pick and shovel, and built himself a cottage, sunk a shaft, and finding ore had it assayed. Some others became interested with him, but after a brief time he sold his interests to Messrs. Hardy & Meader for $65,000; this ground is now owned by the Union Company. . . ."

5. *Mining and Scientific Press*, August 17, 1863.

6. *Ibid.*, August 24, October 26, 1863.

7. By 1857 yield of the Lake Superior region mines was about 5,000 tons of copper annually, valued at $2,000,000. *Ibid.*, April 1, 1865.

8. *Ibid.*, January 12, August 17, 1863.

9. Thomas Price died in San Francisco October 13, 1912, aged 77 years. Professor Price, as he was called, was widely known thoughout the West as a mining engineer and metallurgist, having arrived in San Francisco from Swansea, Wales, in 1862. He led an active career, his professional duties taking him, at one time or another, into all the Pacific Coast states and territories; to Colorado, Montana and Mexico, and even as far east as North Carolina. In 1864 he erected at Antioch, California, the first copper smelting plant in the West. The first plant of the Selby Smelting & Lead Company was built from his plans and under his direction, his Swansea experience giving him a knowledge of the metallurgy of lead, which at that time was unique on the Pacific Coast. The hydrated borate of lime known as "priceite," was first shown by him to be a new mineral, and was so named by Professor Sullivan of Yale, who confirmed Professor Price's original work. *Engineering and Mining Journal*, November 9, 1912, p. 902.

10. *Mining and Scientific Press*, March 16, 1863.

11. L. S. Austin, "Fifty Years of Smelting in the West," *Ibid.*, May 21, 1910, p. 754.

12. *Ibid.*, March 16, July 20, 1863.

13. *Ibid.*, October 26, 1863.

14. *Ibid.*, January 9, July 9, 1864; April 22, 1865; L. S. Austin, "Fifty Years of Smelting in the West," *Ibid.*, May 21, 1910, p. 754.

15. *Ibid.*, November 2, 1863; January 1, April 22, 1865.

16. Paul D'Deirry's, "Metallurgy of Copper in California," appeared in seven installments in the *Mining and Scientific Press* between January 28 and March 11, 1865. Because heavy firebrick was "very dear in San Francisco," D'Heirry recommended manufacturing bricks from local deposits of fire clay. If a suitable clay could not be found, silicious standstone, schist, gneiss, and some granites could be used instead. "A trial by fire" was the best mode of testing these rocks." Once this material was collected a cupola or blast furnace could be erected capable of reducing six tons of ore a day, yielding twenty per cent rosette copper. The cupola was constructed on the outside of stone or ordinary bricks, and lined in the interior with some sort of fire resistent materials, either firebricks or stone, and surmounted with a chimney. A "blasting machine," nothing more than a cylinder and plunger driven by a steam engine, furnished the blast, which was directed through three nozzles (tuyeres): one on either side of the cupola, and one at the back.

"The tuyere should be opened in the backwall," wrote D'Heirry. "The height from the tuyere to the mouth, should be five feet, five inches. The length from the forewall to the tuyere, thirty-one inches; the width from one side to the other twenty-six inches. The forewall has a thickness of only four and a half inches, made of materials in the nature of fire-bricks. The opening for throwing in the ore and coal, above the forewall, is forty-nine inches in breadth and thirty-nine inches in height." The crucible, below the forewall, was twelve inches in depth, and made of *brasque*, a mixture of equal parts of quartz sand, clay and charcoal. The crucible extended six inches in front of the exterior face of the forewall, and ended in a platform made of *brasque*. D'Heirry called this the fore-crucible. A tap hole was made at the lower part of the crucible, and opened in front of the platform, where the basin received molten metal. The tap hole was shut by a plug of clay which was broken at each run of the metal by the aid of an iron bar. The entire furnace was constructed upon a foundation of brick or stone, in which canals were laid to get rid of humidity. The fastenings were of iron. Furnace construction was not peculiar; its description being found in all the treatises on metallurgy. Those who wish to produce only regulus or matte needed only one cupola furnace, stated D'Heirry. Production of black copper required two cupola furnaces; for rosette copper, a small German hearth should be added. D'Heirry estimated $10,000 would cover the cost of erecting two cupola furnaces, a small German hearth, crusher, assorted machinery and implements.

17. *Mining and Scientific Press*, November 2, 1863; December 16, 1865.

18. *Ibid.*, February 17, 1866.

19. *California Farmer*, August 21, 1863.

20. *Mining and Scientific Press*, January 2, 1864.

21. *Ibid.*, September 17, October 29, December 3, 1864.

22. *Ibid.*, April 1, 1865. Desormeaux's process was similar to that described by D'Heirry. The ore was first subjected to calcination to drive off sulphur, a process performed in the open, the heaps being covered with a coating of clay, after the manner of covering a brick kiln.

23. Reproduced in *Ibid.*, September 30, 1865.

24. *Ibid.*, November 23, 1863; March 26, 1864.

25. *Engineering and Mining Journal*, December 4, 1886, pp. 400-401.

26. *Mining and Scientific Press*, March 10, 1866.

27. *Ibid.*, May 14, 1864; November 18, 1865.

28. The smelter described by Piggot was the *fourneau a manche* or French smelter. "Its base is made of very solid masonry, strengthened by transverse bars of iron. The cavity is lined with a coating of very refractory material, cemented to the masonry in the form of an ellipse, and renewed every season. The two lateral faces are constructed of gneiss, and the front, called *fiervende*, of rectangular iron plates, coated with fire-clay. The cavity is a rectangular parallelopiped about six feet long, five and a quarter broad, and three and a quarter deep. The sole is formed of a fire-brick made from Bourgogne clay and pulverized quartz. The tuyere has an orifice three inches in diameter, its muzzle being wrought and its bed cast iron. Opposite it is a platform made of clay firmly beaten, in which there is dug a crucible on a level with the sole of the furnace. It is coated with a mixture of clay and finely powdered charcoal, and from it passes an inclined canal leading to the receiver in which the fluid collects." A. Snowden Piggot, *The Chemistry and Metallurgy of Copper*. Philadelphia: Lindsay and Blakiston, 1858, pp. 337-339; 342, 443.

29. *Mining and Scientific Press*, December 3, 1864; January 21, 1865; November 18, 1865.

30. *Ibid.*, January 6, February 17, April 21, 28, 1866.

31. The Buchanan lead was touted as being quite rich: "At a depth of sixty feet, the vein of ore is five feet wide, and the mineral is of extraordinary rich quality."

32. *Ibid.*, April 7, May 12, 1866.

33. *Ibid.*, January 27, April 7, 21, 1866.

34. *Ibid.*, March 12, 1864; June 6, September 1, 8, October 20, 1866; May 4, 1867.

35. *Engineering and Mining Journal*, January 2, 1886, p. 2. For discussion of who was the first to apply a water-jacket to a smelting furnace in America see *Engineering and Mining Journal*, October 10 (p. 214), October 24 (p. 287), November 14, 1885 (p. 335).

Chapter Two

CERRO GORDO

The next attempt at smelting on the West Coast occurred near Lone Pine, on the west side of Owens Lake, in what would become Inyo County, California. This time the reduction technique was applied not to copper but to argentiferous lead ores. That story started in spring of 1865 when a small group of Hispanics from Nevada prospected the mountains about Lone Pine. On a peak called Cerro Gordo (Fat Mountain), on the western slope of the Coso Mountains, a high range situated 20 miles east of the Sierra Nevada, they discovered a number of argentiferous lead deposits.[1] In early 1866 José Ochoa located the San Lucas claim and began to mine in a systematic manner.[2] Word of the discovery spread, resulting in "quite an immigration of other Mexicans," perhaps some Chilians, and certainly a few Anglo-Americans, and mining took off. On April 6, 1866, the Lone Pine Mining District was organized, encompassing Cerro Gordo.[3]

Staking the best mineral-bearing land, these miners grubbed out the argentiferous lead ore. Some of these men, probably Mexicans, had a sound knowledge of metallurgy. Prior to coming to Nevada, between 1858 and 1865 they had smelted silver-lead ore at the Heintzelman Mine near Arivaca, in what would become Arizona Territory, and in several metalliferous localities along the Colorado River in the area south of Williams' Fork. They recognized that the bulk of the minerals at Cerro Gordo were galena (a lead sulphide), cerrusite and various carbonates of lead; all self-fluxing, easy to melt. They could reduce the ore by smelting, as was done in Sonora and Chihuahua.

They constructed two types of furnaces common to northern Mexico: the *vaso* and the *horno de fundicion*. The latter smelter, actually a shaft furnace popularly called a Catalonian forge,[4] was built in pairs for purposes of strength, side by side, against an adobe wall, under a vaulted roof, 30 to 40 feet high, like that of the mediaeval kitchens, which served to carry away all fumes. Generally they were lined on the back and sides with sandstone, trachyte, porphyry, or other refractory stone, the front wall, however, being usually built of nine-inch thick adobes, laced longitudinally.

Truly a poor man's smelter, the Catalonian or "Mexican furnace" was quickly and cheaply built; no more than 350 bricks going into its construction. The crucible, the front wall together with a tap-hole, lay on a bench consisting of a wall of stone filled with rammed earth or *brasque*, composed of equal parts of charcoal and the most refractory clay obtainable. At the level of the tuyere the furnace generally had an internal cross section 12 inches in depth by 9 inches in width; the height of the furnace above the tuyeres being about 5 feet 6 inches. A small, square chimney, 10 feet high, carried off the fumes.

The blast was furnished by 46-inch bellows, worked by hand and generally of one-man power. Often the bellows were actuated by cams fixed in a horizontal axle connected by toothed gearing with a vertical shaft, to which a mule was attached. The blast was conducted through a slanting hole resembling an auger hole placed on one side of the chimney. In northern Mexico cast copper tuyeres were used, so heavy that they would not be melted by the blast. In this two nozzles of the double hand bellows rested. It is doubtful copper tuyeres were used at Cerro Gordo.

The Catalonian furnace, which had a capacity of about five tons of ore per day, was fed through a small hole about five feet from the ground; a quart of ore and two quarts of charcoal being thrown in alternately until the chimney was full. Fire was then applied, the bellows set in motion, and the furnace brought to operating temperature. The mechanism was nothing more than a giant blow-pipe. An iron bar, to skim the slag from the metal, was all that was needed. The molted metal ran out a hole on the opposite side of the furnace directly onto the ground or was collected in clay molds. The Catalonian furnace was cheap to run; the daily cost being no more than $30 for charcoal, wood and labor.

The second type of furnace, the *vaso* or hearth, could also be used to reduce lead ore to bullion, providing it was built with a steeply inclined hearth. In Chihuahua and Sonora such a furnace was called a

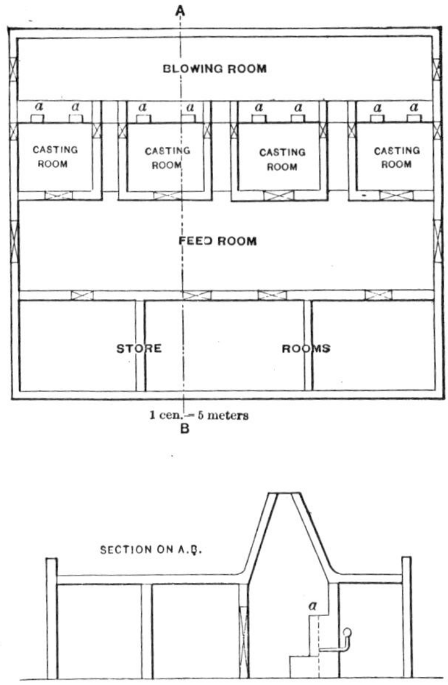

Fig. 1.—PLAN AND SECTION OF MEXICAN SMELTING WORKS.

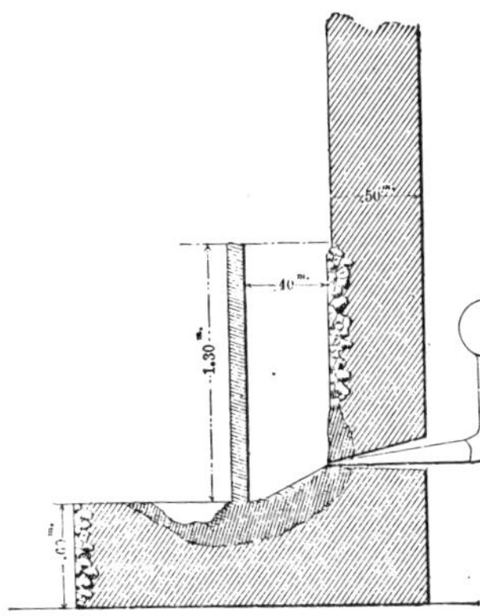

Fig. 2.—SECTION OF UPRIGHT FURNACE.

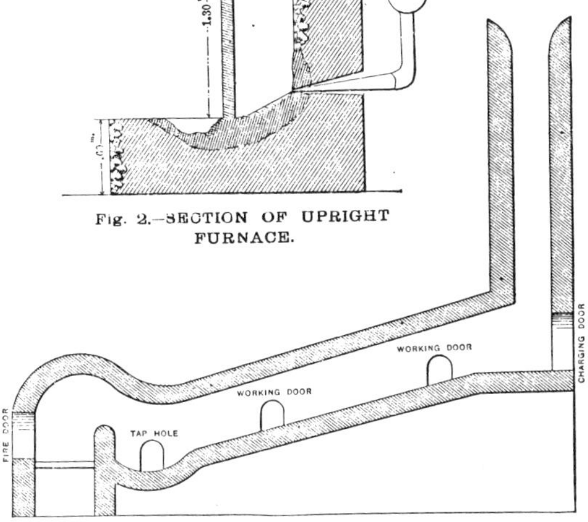

Fig. 3.—SECTION OF SERPENTON FURNACE.

Top: Plan and section of a Mexican smelting works. Section of upright furnace, *horno castellano,* in middle. The reverberatory *vaso* or serpenton furnace at bottom. Of the two, the *vaso* was more durable and economical; far superior to the upright furnace for smelting pure carbonates and galena. (From *Mining and Scientific Press,* July 30, 1887)

serpenton. With slight modification it could be utilized to separate silver from lead bullion produced in shaft furnaces. This process, called cupellation, had long been used by assayers to separate gold, silver, or other nonoxidizing metals from base metals which can be oxidized. That was done by placing the metallic mixture in a cupel, a shallow, porous cup, and roasting in a blast of air. The base metal oxides are absorbed in the cupel, leaving pure metal to be decanted. The process was carried on in a much larger scale by Mexican miners at Cerro Gordo, using the *vaso*.

In contrast to the Catalonian shaft smelter, the *vaso* was a reverberatory smelter. It was the epitome of simplicity; required no expensive metal castings, no refractory building materials, and was cheap to operate, the fuel being small sticks of scrub-oak, or any kind of brush. The *vaso* could be built in 24 hours at a cost of no more than $15.

The foundation of the *vaso* was about 18 inches thick, and rested between adobe sidewalls. A six by four foot hearth, composed of a mixture of two to three parts hardwood ashes to one part of clay, well rammed, was laid between the sidewalls. The fireplace was 18 inches wide and six feet long. A shallow basin, the equivalent to the cupel, was cut out from the hearth. The roof of the furnace was not arched, but built simply of adobes of extra large size (27 to 30 inches long, 11 to 12 inches wide, and 3½ inches thick). It rested upon bricks standing on edge, which formed the sidewalls. All joints between the adobes were carefully luted with clay. Blast was supplied from an ordinary forge bellows through a single tuyere positioned just above the basin. The furnace had no chimney or flue-chamber.

In firing up the *vaso*, care had to be taken to bring the furnace slowly to operating temperature so that cracks did not develop. A test run was always conducted to assure there was no cracking. If cracks appeared, six or eight inches of the surface was chiseled away and replaced with fresh well-tamped material. Further tests were conducted until the furnace was deemed free of cracks.

The furnace was heated for about 18 hours, and then charged with about 300 pounds of bullion. When full the bellows were pumped to supply the equivalent of a blast. As crude as it was, the *vaso* worked. As the bullion melted, more was added. Litharge or dross was allowed to run from a notch opposite the tuyere. More bullion was added at intervals and melted down. During course of operation, the lead, the bulk of the bullion, was oxidized, the silver collecting in the shallow basin. When the mass of precious metal amounted to 80 to 100 pounds, the

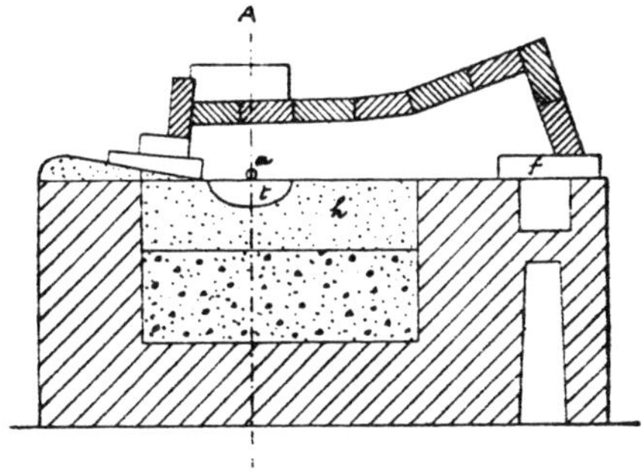

FIG. 9.—Longitudinal Section through C D.

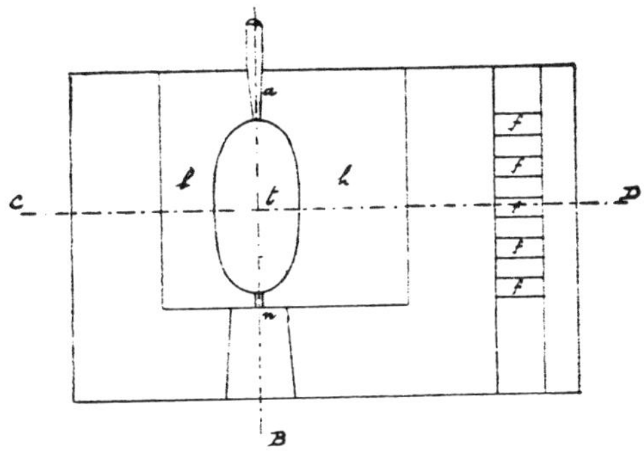

FIG. 10.—Plan on Tuyere Level.

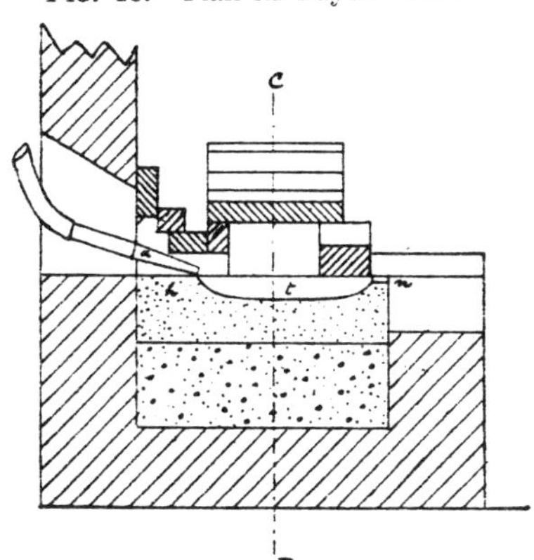

FIG. 11.—Transverse Section through A B.

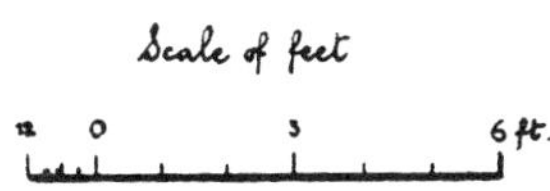

CUPELLATION HEARTH.

The adobe *vaso* or cupellation furnace.

furnace was cooled sufficiently for it to set; an adobe was removed from the flat roof, and the cake of silver, .996 fine, lifted out and cooled in water. The basin was patched, the roof repaired, and another furnace run, called a campaign, commenced. The entire operation was conducted by two men per shift.[5]

Although Hispanic miners produced "quite a quantity of bullion" in their shaft furnaces, and separated the silver from the lead in *vasos*, they had difficulty disposing of the precious metal because it could not be "sold for coin." In 1867 coin was unavailable in the area, and there were no assayers in the region to ascertain exact values. This compelled miners to trade off their silver for merchandise and provisions at about $1 per ounce." Consequently, hardworking miners were barely able to subsist, which set the stage for the entrance of middlemen who grabbed off valuable mining properties.

While miners at Cerro Gordo lived from hand-to-mouth, their region was being eyed by Anglo-American mining men. Among the first to arrive was "Doc" James Delavan. Why he was called Doc no one knows. He never practiced medicine and so far as the West is concerned he was always involved in mining. Born in Ovid, Seneca County, New York, on July 9, 1815, Delavan arrived at San Francisco aboard the steamer *Panama* on April 1, 1849. He mined gold in California for more than a decade and then switched his operations to Washoe. Among the first to pursue mining on the Comstock, he attained prominence at Virginia City as superintendent of the Mount Davidson Tunnel and the Lady Bryan Mine. In fall of 1867 he moved to Cerro Gordo.

As representative of the newly organized Cerro Gordo Mining Company, which also went under the name of the Virginia Company, Delavan secured claims from original locators at prices ranging from fifty cents to $20 per foot, and began mining. In winter 1867 he constructed a smelter, a shaft furnace and a reverberatory hearth, both not much better in design than those constructed by Mexicans. In Nevada and San Francisco newspapers Delavan hyped the area. "I know of no place on the Pacific Coast that holds out so great an inducement as this for persons of capital and enterprise, and there will probably be a large emigration here next season."[6]

Delavan operated less than a year at Cerro Gordo, all the time hoping to interest entrepreneurs. In the end his smelter failed, and most mining men regarded his dispatches from the Lone Pine region as so much hyperbole. They had heard it before; but not all turned a deaf ear. On April 2, 1868, Delavan reported in the Virginia City

Mortimer William Belshaw

Territorial Enterprise that "parties have already arrived in the camp from San Francisco, have secured mines, and are about erecting furnaces."[7] The "parties" from San Francisco were Victor Beaudry, Mortimer Belshaw, and Abner Elder. Other than possessing shrewd business sense, Belshaw and Beaudry had little in common. But they would dominate the Cerro Gordo scene for the next decade.

Described as a "practical engineer," Mortimer William Belshaw arrived at Cerro Gordo in April 1868 to investigate the area's silver-lead deposits. He was of Irish descent, his great grandparents having emigrated from County Antrim, Ireland, to New Haven, Connecticut, about 1770. The family later moved to central New York, where Mortimer was born in 1830. Educated in local schools, Mort as he was called, worked in a jewelry shop, and graduated from Geneva College. He collected tolls on the Erie Canal until smitten with gold fever in 1852. Obtaining passage on a steamer to Panama, he crossed the Isthmus and proceeded to San Francisco. At 25 years of age, Belshaw was operating a jewelry store in Fiddletown, buying gold dust, and serving as an agent for Wells, Fargo Express. In 1858 he returned to New York, married his sweetheart, Jenny Oxner, and brought her back to California.[8]

In 1862 Belshaw went to Mexico, and for several years operated a silver mine in Sinaloa that may have been financed by San Francisco capitalists. Political instability south of the border forced his return to San Francisco. There, he established the Pacific Refining and Bullion Exchange. Thus, Belshaw was aware of the money-making potential at Cerro Gordo before its mines made newspaper headlines. Enlisting Abner Elder, another veteran of Mexican silver mining, Belshaw headed for Cerro Gordo in early 1868. They inspected claims, poked around adits, and examined primitive smelters. On May 6 Belshaw and Elder obtained a third interest in the Union Mine from Joaquin Almada. They mined a few tons of ore, and rented a *vaso* to process it. The two men, however, had grander plans. Instead of just mining and processing their ore, they saw an opportunity to make lots of money running what was commonly called a custom smelter, refining not only their ore but that of other mines. To do that they would need an expensive large capacity smelter. But they knew how to raise money.

Belshaw and Elder freighted 3,500 pounds of bullion to Los Angeles in June 1868, and accompanied the shipment to San Francisco. The bullion was refined and the two men used the silver to intice investors. They didn't have to work hard to draw the attention of capitalists. Egbert Judson, one of San Francisco's distinguished millionaires, and a promoter of industrial enterprise on the Pacific Coast, joined the Cerro Gordo venture.

Born about 1812 in Syracuse, New York, Egbert Judson showed up in California in 1850. He went immediately to the mines, but after a year he went East. He returned to San Francisco in 1852 to establish an assay office. In 1855 he became interested in the manufacture of acid and started the San Francisco Chemical Works. He also invested in placer mines in Nevada and Butte counties, becoming one of the principal owners in the North Bloomfield mines, the largest hydraulic operation in the state. He also owned the Kennedy Mine in Amador County, one of the state's leading gold-producers. Judson's name, however, is not so much linked to mining as it is to other enterprises.

By 1868 the name Judson was synonomous with San Francisco's blasting powder business. Obtaining a license from Alfred Nobel to manufacture dynamite, Judson formed the Giant Powder Company, the first to manufacture high explosives on the Pacific Coast. He also invented and patented Judson powder for bank-blasting, a substance which soon displaced others in gravel-mining. The original high explosives were composed of nitroglycerin and absorbents; but his powder consisted of particles of black powder rendered nonabsorbent by a peculiar process, mixed with nitroglycerin so that the nitroglycerin remained on the surface. Only a small proportion of nitroglycerin was used, but the powder was effective. Sales of the new product were immense, and the substance was patented all over the world. Judson later organized the Judson Dynamite and Powder Company and built extensive works on the bay shore in Alameda County, the largest powder works at the time on the coast. In forming this company he associated with some of the most prominent men and largest mine owners in the state, all directly interested in development of the mining industry. Their idea of joining together to manufacture blasting powder was the result of their determination to avoid payment of high rates previously charged for explosives. The company achieved its goal, forcing down the price of powder from 18 cents to 12 cents a pound.

Judson ventures were many: the Judson Manufacturing Company, Judson Dynamite and Powder Company, California Paper Company, Western Fuse and Explosives Company, San Francisco Candle Works; San Francisco Chemical Works, California Pulp Works, Sather Bank, Pioneer Pulp Works, and a chemical works at Newark, New Jersey. Because he made enormous sums of money.[9] Judson was the man to have in any mining venture.

With Judson as a silent partner, the Union Mining Company was organized, and Belshaw and Elder returned to Cerro Gordo with capital in hand. This deal, of course, created strange bedfellows. Belshaw, Elder, and Judson were now linked in business with Victor Beaudry, who owned 50 percent of the Union Mine. Flushed with money this partnership moved quickly to dominate mining and smelting at Cerro Gordo.

In summer of 1868 Belshaw and Elder graded a wagon road from Owens Lake to the upper end of the town of Cerro Gordo. They carted in firebrick, cast iron, and a boiler and steam engine. A crew of masons and mechanics followed, and within weeks a large vertical

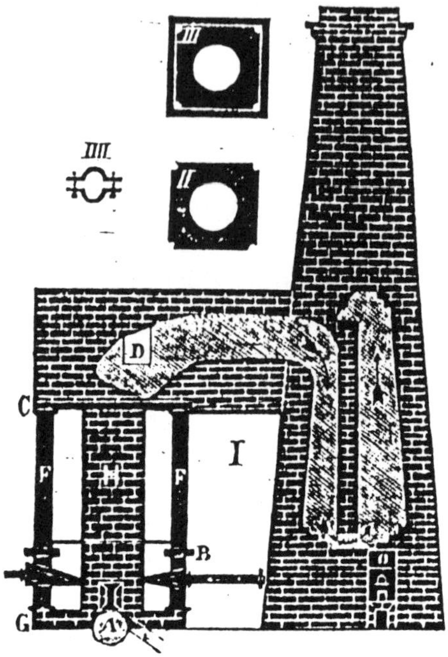

Diagram of Belshaw's first furnace at Cerro Gordo as depicted in the *Mining and Scientific Press* of March 4, 1871. The "hood" (Fig. I) of red brick rests upon a diaphragm, C, shown in Fig. III, which rests on four cast iron pillars, F. At a distance of three feet above the foundation, G, was a supplemental diaphragm (Fig. II), which supports the upper portion of the furnace, H, between the two diaphragms. The supplemental diaphragm could be raised or lowered by means of collars, Fig. III, placed around the pillars at B.

blast or cupola furnace, with a square horizontal mid-section, took shape. The hearth or crucible was made of stone, the front part of which was formed of *brasque,* a composite of charcoal and fire clay. At the bottom of the hearth was a tap-hole to draw off the molted lead.

Above the hearth were two sections made of firebrick. The first, called the tuyere section, two feet above the hearth, was four feet high, and about four feet, six inches long by three feet wide. This section contained two cast iron tuyeres or nozzles which provided the blast. That number could be increased to five. At each corner of this section stood a cast iron pillar which supported the second part

Victor Beaudry

of the furnace, an eight foot high superstructure, which commenced four feet above the tuyeres. The iron supports facilitated easy repair of the crucible and tuyeres, components most subject to damage by heat.

The shaft of the furnace (the hollow section where fusion occurred) was circular, 20 inches wide at the crucible, with an inside batter of 95 degrees to the top of the superstructure. This superstructure was built of radial firebrick (14 inches long), moulded in the form to make the inside diameter of the shaft. Outside circumference of the superstructure was 12 feet, and was hooded over all the way to the top with common red brick. A square hole, 18 by 24 inches, was provided in the side of the superstructure to admit the charges. Thus, the furnace was rectangular in shape, the interior space being circular. A pressure blower, probably a No. 3 Sturtevant fan, three feet in diameter, driven by a 10-horsepower steam engine, supplied the blast; the air conveyed to the tuyeres through canvas pipes.

Unlike Haskell's copper smelter, Belshaw's first furnace did not have a water-jacket. The reason for that was simple. Water at Cerro Gordo was in short supply, costing ten cents a gallon. Imported fire-brick and thick wall construction was deemed sufficient to withstand the heat. The furnace tuyeres, however, were water cooled, Elder making the mould himself for what he called the "water tuyere." It "worked well," he reported years later.[10]

Tens of thousands of bushels of charcoal were brought from Truckee, and by end of September the furnace was ready. Before it could be blown-in furnace walls and crucible had to properly cured. This was done by bringing the smelter slowly to operating temperature; then it was loaded with ore and charcoal, and the blast applied. The level of the charge was kept as low as possible without having a flaming top. Smoke and sulphurous gases belched from the chimney. As the smelter had no flues, large quantities of dust, laden with lead, were blown out of the shaft to settle about the works. In three days 15 tons of ore was fed through the smelter. The yield was eight tons of silver-lead bullion. In seven days 385 bars, weighing 18 tons, containing an average of $250 per ton in silver, were produced. When the bullion was shipped to San Francisco at a cost of four cents per pound, the *Mining and Scientific Press* reported "there is no longer any question of the success of smelting operation here."[11]

While Belshaw and Elder were getting their smelter going, Victor Beaudry was hard at work. Like James Delavan, this entrepreneur was early on the California scene. The similarity stops there. A French Canadian, Victor came from a prominent Quebec merchant family. He arrived on the Pacific Coast in 1849 with his brother Prudent. While Victor remained in San Francisco to engaged in commerce, Prudent moved south to Los Angeles to speculate in real estate.

Victor established a shipping business on the Sacramento River, and together, the Beaudry brothers organized a line of sailing vessels running between Panama and California carrying passengers and freight. In 1851 Victor went to Panama and thence to Toboga, Nicaragua, where he amassed a fortune in a short time. In 1855 he returned to California to partner with brother Prudent in various ventures. Joining the rush in early 1861 to work gold placers in San Gabriel Canyon, Victor was elected president of that new mining district. At outbreak of the Civil War, when troops were stationed in Los Angeles, Beaudry became sutler for the First Infantry Regiment. In late 1861 he opened another sutler's store in San Bernardino, and when troops were detailed to subdue Indians in Inyo County, he opened another outlet at Camp Independence in 1866. That year he launched a general merchandising venture at Cerro Gordo. A hard-nosed capitalist, Beaudry was in the right place at the right time.

From his Owens Valley and Cerro Gordo stores Victor extended liberal credit to miners. As mentioned, unable to sell their bullion and silver, the miners were cash strapped. A perfect opportunity for Victor. When they defaulted on their debts, Beaudry foreclosed. Winning judgments he gained the San Felipe quartz ledge, the Merced and the Ygnacio claims, and a half interest in the district's principal producer, the Union Mine. Thus he and Mortimer Belshaw became partners.

Mining property is worthless without a means to process the ore. Knowing that smelting facilities at Cerro Gordo were inadequate, Beaudry moved to build a smelter large enough to handle ore from his properties as well as other mines. He sought talent to insure ore reduction techniques meshed with Cerro Gordo mineralogy. Pierre Desormeaux, a metallurgist with extensive experience in smelting base metals, was employed to design an up-to-date refinery.

Inspecting Beaudry's Cerro Gordo holdings, Pierre recognized two types of ore: a rich silver and gold bearing quartz, which Mexican miners avoided, and a vast amount of lead sulphide or galena, previously used as flux for richer ores, and lesser amounts of lead carbonates carrying high percentages of silver. By what he called "scorification," a double fusion, Desormeaux proposed to exploit the vast deposits of galena.[12]

A disciple of Paul D'Heirry, this Frenchman was a practitioner of the French method of smelting. As mentioned earlier, in fall 1864 Desormeaux erected two blast furnaces at the Union Mine at Copperopolis. When those smelters burned in October 1865, he replaced them with smelters of newer design. On January 24, 1868, Beaudry contracted with Pierre Desormeaux to construct a refinery close to his mines.[13]

Desormeaux erected two furnaces: a reverberatory and a shaft furnace. The reverberatory was large, 22 feet long by six feet wide. Its interior was lined with *chemise* or firebrick. In this smelter, "of his own design," the Frenchman adopted a double fusion technique. After the furnace was thoroughly dryed, it was charged with 1½ tons of galena from the Union Mine. Well spread on the hearth, the ore was scorified for four hours to drive off the sulphur and arsenic. Silver-bear quartz was added, equal to about a third of the original charge. The heat was raised and maintained for two hours until complete fusion had occurred, to produce what Desormeaux called "agglomeration." The result was a lead matte, rich in silver.

This matte was handled in simple fashion; run onto the ground, allowed to cool, and then broken up and piled away to be used again in the blast furnace. In a 15 day and night campaign, Pierre "scorified" 44 tons of galena and nine tons of other ore, to produce 35 bars of bullion, weighing 1½ tons, and containing an average of $500 per ton in silver. His reverberatory was not "the least injured" by the run, and Desmoreaux pronounced his smelter a success.[14]

By subjecting the matte to a second fusion more impurities could be driven off, and the silver content raised. That process was accomplished in a circular blast furnace, about eight feet in height and 20 inches in diameter (on the inside) at the bottom, with an inside batter of 95° to the top of the eight feet; identical to Belshaw's furnace. The shaft furnace ran from 12 to 15 days, charged with five parts matte or slag; one part raw galena ore, broken to egg size; one part raw silver-quartz, broken rather smaller; two parts galena and iron oxides; and one part carbonate of lime, with about two bushels of charcoal.

Richness of the charges were regulated by decreasing or increasing the constituents. Molten bullion, carrying lead, silver, gold, and other metals not drossed off, ran into a receptacle in front of, and on a level with the bottom of the furnace. Slag also ran into this receptacle, overflowed, and ran off. When cooled it was wheeled away and dumped into a pile as of no further use. The molten bullion in the receptacle in front of the furnace was tapped at the bottom, and allowed to flow into a hole or pot, made of fire clay sunk into the earthen floor of the building. There it is permitted to stand a while to collect "*fierros*" or a very rich matte containing silver, copper, iron and antimony, which formed a hard scum at the top of the pot, varying in thickness from half an inch to three inches. This scum was removed and the lead ladled into iron moulds, forming pigs 83 pounds in weight. When running well, the furnace produced 80 pigs in 24 hours. Desmoreaux's process was just the reverse of that used by Mexican miners, and the bullion was not subject to cupellation.[15]

The furnaces ran well and by August 1868 Beaudry's operation was in full blast. A furnace run in September produced 104 bars of bullion, weighing 8,300 pounds, worth $3,864. A month later, however, Desormeaux transferred to the smelting operation of Belshaw, Elder, and Judson. By year's end the two furnaces were working not only ore from the Union Mine, but ore from the San Ignacio, Belmont and Pedro mines, each furnace producing an incredible four tons of bullion a day.[16]

In early January 1870, 35,000 pounds of bullion was shipped to Los Angeles. Seven tons more arrived in Los Angeles on February 5, and there were three trains upon the road with 15,000 pounds of bullion, and 20,000 pounds more were being hauled by ox teams. Thirty tons of bullion were awaiting shipment at Cerro Gordo. Obviously, the two furnaces were producing bullion faster than it could be hauled away.[17]

Not only did Belshaw and Beaudry control the principal mines, they monopolized smelting at Cerro Gordo, and a friendly rivalry existed between their furnaces. Writing in the *Mining and Scientific Press of* May 21, 1910, Francis Drake stated that "the ore as it came from the mines was sorted, the galena and waste being picked out; it was then divided on the dump. One wagon load went to the furnace owned by Messrs. Judson and Belshaw, and the next went to the furnace owned by Mr. Beaudry, and so in turn the galena was weighed, but no trouble was taken with the ordinary ore. I remember Mr. Belshaw defending the separate treatment of ore by saying that the rivalry between the two furnaces treating the same ore tended to make those in control put forth every effort. There certainly was rivalry. One of my duties was to try to find out the size of the other fellow's bullion pile, and if either furnaces was shut down longer than usual for the almost daily patching of its brick walls, there was gratification in the rival office.

"No attempt was made to weigh the quantities of materials going into the furnace. The fuel was charcoal and the quantity to each charge was a 'box' of indefinite capacity as it had only three sides and was unequally filled by the charcoal passer whose work was to fill the box from the charcoal heap and inject its contents into the furnace when told to do so by the feeder. The superintendent exercised some supervision over the number of shovelfuls of galena, quartz, and slag to each charge, also told the ore-wheeler what proportion of the different colored ores to wheel on to the charge floor; but the feeder put in as many shovelfuls of this ore mixture as he thought the furnace would stand and keep running." Weighing of the furnace charge was introduced at Cerro Gordo in 1878 by Fred F. Thomas.

While both smelters were declared successes, there were problems. Because they were neither water or air cooled, expensive firebrick and mortar used in construction melted away. Fire clay crucibles stood only 30 days. Imported firebricks were replaced with bricks made of kaolin found nearby. Mixed with one-third pure white quartz sand, this kaolin produced an excellent fire-resistent brick. Not burned, the bricks served as a flux for highly basic ores then being treated. That was only a minor irritant compared to the problem of outside competition.

Others flocked to Inyo County to get into the Cerro Gordo smelting act. A group of New Yorkers, headed by Joseph Darrah, organized the Owens Lake Silver Smelting Company, and under superintendency of James Brady, installed a smelter on the eastern shore of Owens Lake, about eight miles from Cerro Gordo. This refinery consisted of two reverberatory furnaces and one blast furnace; the whole driven by a 40-horsepower engine. An artesian well, sunk in June 1870, provided ample water for the boiler. The ore was freighted from Cerro Gordo mines; cord wood and charcoal from the Sierra Nevada were ferried across the lake on flatboats from Cottonwood Canyon, a distance of 15 miles. The company's plant was running by January 1870 and in May it shipped 2,000 pigs of bullion to Los Angeles.

That spring Wolfskill & Company of Los Angeles erected a large smelting works, run by water power, two miles west of Lone Pine, as well as furnaces on Cottonwood Creek on the west beach of Owens Lake. At the same time, the Eclipse Mining Company, an English-owned firm managed by Henry Tregellas, a mining engineer with experience in England, Australia, and California, purchased the property of the Union Mill and Mining Company for $150,000. While they had a mill on Owens River, the company erected a furnace in early 1871 to process argentiferous galena.[18]

Belshaw and Beaudry effectively handled that competition. They owned the Union Mine, the principal lead-producer, and all traffic between Owens Lake and Cerro Gordo had to pass over their toll

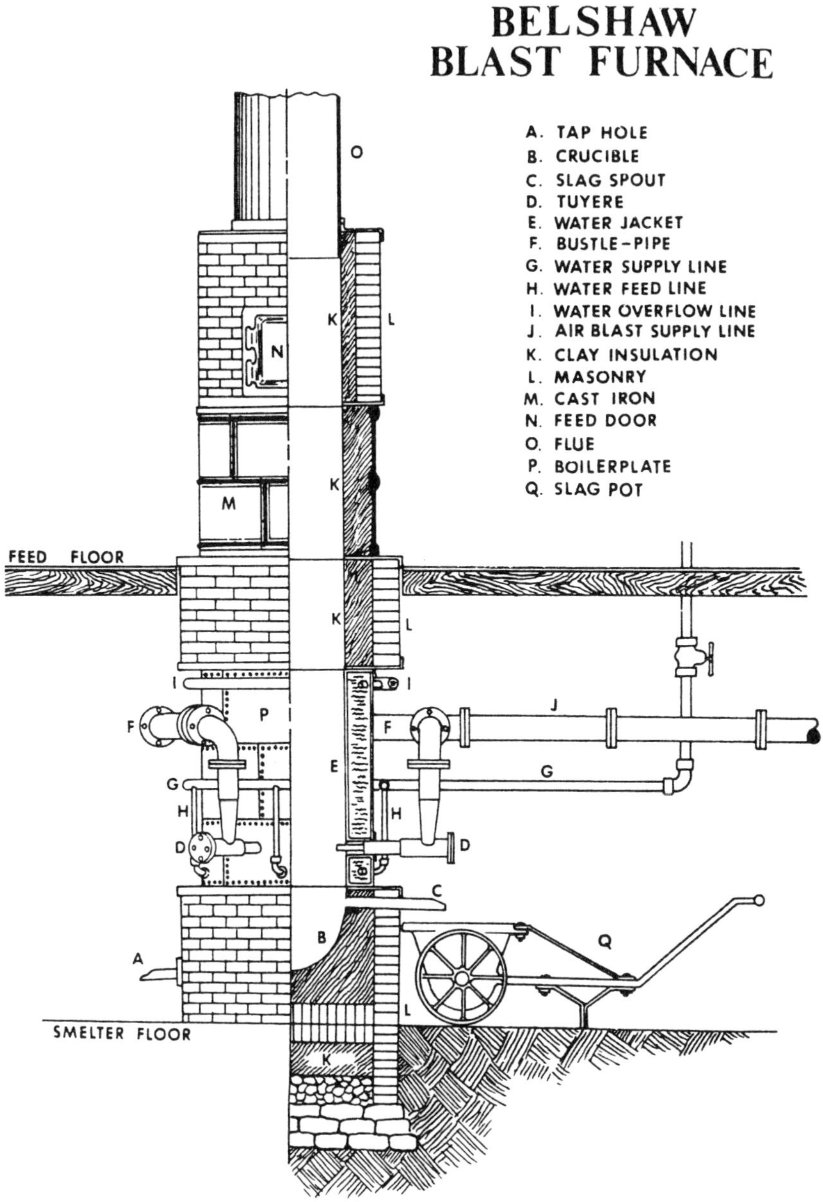

Diagram of Belshaw's furnace following the 1874 installation of water-jacket.
(Courtesy Don Chaput)

road. Cerro Gordo was their monopoly. Nevertheless logistical problems were horrific.

The Cerro Gordo mines at were at an elevation of nearly 8,200 feet above sea level; a mile above the floor of Owens Valley. The valley was remote and arid. There was a serious lack of water which had to be remedied if smelting was to continue. Either water had to be carried to the ore, or the ore carried to water. So far as the Mexican smelters were concerned there was no problem obtaining fuel. Vasos could be stoked with brush, scrub oak, and mountain mahogany — growth that could also be converted to charcoal for shaft furnaces.

The blast furnaces of Belshaw and Beaudry were voracious consumers of charcoal, and it was not long before nearby mountain slopes and valley floors were denuded of growth. Until 1872, when kilns were erected at Cottonwood Canyon, charcoal had to be hauled in from Truckee and ferried across Owens Lake. Wood for building purposes was cut in forests on the eastern scarp of the Sierra Nevada, 30 miles distant, floated down streams to Owens Valley and hauled eight miles to Cerro Gordo.

Transportation was by far the most pressing problem. The ultimate destination of Cerro Gordo bullion were San Francisco refineries and brokers. Between Cerro Gordo and San Francisco stood the Sierra Nevada. The nearest passes were hundred of miles away and closed by snow in winter. The only practical alternative for Belshaw and Beaudry was to skirt the mountains to the south, take the ingots by wagon to Los Angeles, and from there ship the bullion to San Francisco. It was an arduous 275 mile journey, a long haul through arid, rough country: lower Owens Valley, Red Rock Canyon, the Mojave Desert, and finally through Cahuenga Pass.

For solution to that problem, Belshaw and Beaudry turned to Remi Nadeau, a veteran teamster who had hauled supplies across the deserts of California and Nevada. If anyone could navigate the arid region, it was this 32-year-old French-Canadian. For several years Nadeau teams carted seven-ton loads of silver-lead ingots into Los Angeles, their "bells jingling, wood and leather creaking, blacksnake (whip) popping, mule-skinner shouting and cursing, mules snorting and coughing."

Despite his satisfactory service, Belshaw and Beaudry did not renew Nadeau's contract in 1871; they had been lured away by competing teamsters. For the next two years transportation of Cerro

Gordo bullion was a hit or miss proposition, which resulted in competition between Los Angeles, Ventura, Visalia, Santa Barbara, and Bakersfield for the Cerro Gordo trade. Visalia and Bakersfield in particular touted the prospects of future rail connections with San Francisco. The stakes were high with considerable commerce involved in transshipment of bullion north and provisioning and maintaining both Cerro Gordo and its wagon system.

In the end teamsters retained by Belshaw and Beaudry proved unreliable. The result was that thousands of bars of bullion backed up at Cerro Gordo and Owens Lake staging points. On top of that mining and smelting equipment broke down. Replacement parts did not arrive on time, resulted in shutdowns. When cash flow was crimped, Belshaw and Beaudry again sought the service of Remi Nadeau.

Assessing the snarled situation, the three men formed the Cerro Gordo Freighting Company and worked out a systematic transportation scheme. A dozen stations, a day's haul apart, each with water, feed, and repair facilities, would be maintained between Cerro Gordo and Los Angeles, the terminus. Eighty teams of 14 mules pulling three wagons would be kept constantly on the road. Averaging eight trips a year, the teams worked like a conveyor belt, hauling bullion southwest to Los Angeles and provisions northeast to Cerro Gordo. [19]

With the transportation dilemma solved and competition neutralized, Belshaw and Beaudry improved their operations by modernizing their smelters, first by increasing the number of tuyeres in each furnace from two to five. Then in 1874 the portion of furnaces which endured the highest temperatures — a section of superstructure extending four feet above the crucible and tuyeres — was redesigned. This portion of the furnace, constructed of large radial bricks made of local fire clay and crushed quartz, was made a foot longer and wider to allow space for a lining made of air-dried bricks half as long as those previously used. Around this lining was fitted a one-piece wrought iron water-jacket. The jacket had large openings for insertion of tuyeres that projected through the jacket about six inches.

It has been claimed that this was the first water-jacket fitted to a lead smelter on the West Coast. It was not. That distinction, as pointed out, goes to Haskell's furnace. By 1874 there were numerous smelters in Nevada utilizing water-jackets. More on that in the

Thomas Selby

next chapter. Nevertheless, those water-jackets were Cerro Gordo innovations.

Designed by Mortimer Belshaw and William E. Goodrum, the latter furnace superintendent at the time, the jacket was made in San Francisco by Hinckley, Spiers & Hayes at their Fulton Iron Works. The jacket, capable of standing high boiler pressure, had a cylindrical steam drum at the back connected by two three-inch pipes (about three feet long) to the steam drum of the boiler. The bottom of the jacket was connected to the bottom of the boiler.

At the same time, in 1874, Beaudry's furnace was also rebuilt, and fitted with a wrought iron water-jacket manufactured by Ira B. Rankin at the Pacific Iron Works. The water-jackets improved the economy of both smelting works, eliminating suspensions of work to patch furnace walls. But faulty metallurgical work and occasional leaks in the jacket caused furnace freeze-ups.

Flues were also added to the furnaces at this time to catch large quantities of dust and lead particles blown from the smelters. The flues were about six feet long and extended downward at an angle of about 30° from one side of the superstructure of the furnace to a brick chimney, which extended a short distance above the roof of the furnace building. The dust was taken out of a door at the base of this chimney.[20]

By 1874 the furnaces of Belshaw and Beaudry were turning out about 13 tons of bullion per day. Considering that this bullion was worth about $115 in silver and $90 in lead per ton, their operation was profitable from the very start; the lead paying expenses of mining, smelting and transportation to San Francisco, where it was refined at a cost of $25 per ton, leaving a net profit of about $90 per ton. The statistics of their operation are staggering. In 1869 a thousand tons of bullion was produced with a net profit of $90,000. Fifteen hundred tons were produced in 1870, netting $135,000. It went up drastically from there: 1871, 2,500 tons bullion produced, net profit $225,000; 1872, 4,000 tons produced, net profit $360,000; 1873, 5,000 tons produced, net profit $450,000; 1874, 6,000 tons produced, net profit $540,000. A correspondent of the *Mining and Scientific Press* wrote that as a camp "Cerro Gordo . . . deserves credit for having supported itself and made money for its miners and furnace owners from the beginning."[21]

The beneficiary of this production was Thomas H. Selby, owner of the San Francisco refinery which separated the silver from the lead and marketed both. Cerro Gordo laid the foundation for his empire, much to the benefit of base metal producers throughout the American West.

Thomas H. Selby was born in New York City about 1820, a descendant of an English family whose history commenced as early as 1069, in the days of William the Conquerer. The American branch of the family settled at Annapolis in the 17th Century, becoming large land owners. During the Revolution the Selbys enjoyed political prominence, and New Yorker John Selby, Thomas's father, gained considerable fame during the War of 1812 for his daring maritime adventures.

Young Thomas was educated in New York City, and at an early age was employed as a clerk in the dry goods house of A. T. Stewart & Company, where he worked alongside of Cyrus W. Field, who became a cable projector and inventor. At age 19 Selby was elected to the position of director of the New York Mercantile Library. Six years later he entered the mercantile business in partnership with B. Naylor, a prominent New Yorker. Under Selby's managment their establishment grew to 40 employees, but in the end the venture failed. Looking for greener pastures, and intending to ultimately pay off his debts, Thomas sailed for California, landing at San Francisco in August 1849.

Selby entered the mercantile business in the Bay City, and did well. In summer of 1850 he erected a large brick building on the north side of California Street, near Montgomery, and originated the firm of Thomas H. Selby & Company. In conjunction with Naylor, his New York partner, he commenced the importation of metals and merchandise from the East.[22]

In August 1865 Selby and Naylor conceived the idea of supplying San Francisco with lead pipe, solder, sheet and bar lead, and the West Coast with bullets and shot of all caliber. They bought a modest refinery and erected a 175 foot high shot tower on the southeast corner of First and Howard streets. Although a prominent and well-known San Francisco landmark, the tower was not an immediate success. Only after heavy expenditure and perseverance did the project become profitable. In the end, the shot tower drove all other manufacturers of shot out of the West Coast market. Manufacture of shot did not establish Selby as much as his refining of bullion.

Before 1865 Selby had not entertained the idea of supplying Eastern markets with lead. That changed, however, with discovery of large deposits of argentiferous lead throughout the West. As there was only one other firm in the West refining lead — the Trinity and Sacramento Silver Mining Company in Humbolt County, Nevada — most high-grade ore and bullion was shipped to smelters in Chicago, Illinois, and Newark, New Jersey, or to the Vivian Works at Swansea, Wales.[23] Opportunity was indeed at hand. Selby could purchase high grade ore or bullion, refine it, and market its constituents not only on the West Coast but in the eastern United States as well.

With backing of prominent investors — C. Flood, W. S. O'Brien, A. J. Ralston, and N. K. Masten — the Selby Smelting and Lead Company was organized in 1867. Land was acquired at Black Point opposite Fort Alcatraz. There, at the northern extremity of San Francisco, the company built a modern refining facility and a bulkhead wharf to accommodate vessels bringing ore and bullion to the works. At the same time, Selby established an agency at the head of navigation on the Colorado River to purchase ore, pig lead and copper matte produced by smelting works erected at the mines.[24]

Unlike the shot business, refining and separation of base and precious metals took off. Production of the Cerro Gordo mines spurred it even more. Production increased from 1,900 tons of lead in 1870 to 12,000 tons in 1874, and the value from $200,000 in 1870 to $1,500,000 in 1874. In 1874 the plant produced 12,000 tons of pig lead, of which 10,000 tons was shipped to New York. The capacity of the plant was not less than 20,000 tons annually, so that all the crude bullion produced on the coast could be refined in San Francisco. Labor employed increased in the same ratio: from 15 hands in 1870 to 70 in 1874.

The secret of Selby's success was that his company bought ore and bullion through agencies established in mining regions. The works always paid shipping expenses on ore from anywhere, promptly meeting freight bills on delivery. Processing was done in San Francisco cheaper and faster than at other refineries in the United States. And Selby's return to miners was fairer than those from Swansea, "honorably and promptly paid."[25]

Selby's financial dealings led to political prominence. In April 1851 he was elected an alderman from the Fifth Ward. In fall of 1852 he was reelected to that office. In 1869 he was elected mayor of San Francisco, and he was a candidate for the governorship of California in 1871, but withdrew his name.

The Selby Company was just beginning to bloom when its founder, Thomas, contracted something diagnosed as inflammation of the lungs. He died on June 9, 1875, in San Francisco at his Harrison Street residence, leaving a wife, three sons, and four daughters.[26] The company, however, went on under the direction of Prentiss Selby, Thomas's son, who expanded and mechanized the works, making the company the most important lead refinery in the United States.[27]

Refining of base metals meshed with other adjuncts of the Selby Company. In 1876 company products were sheet and bar lead, solder of every variety, and all sizes of lead pipe ranging from one-eighth to six inches in internal diameter. Shot of every size was made. The company also manufactured block tin pipe for dwellings, preferable to expensive tin-lined lead pipe. English tin mines being exhausted, the metal was imported from Australia. Surplus pig lead, not used in

San Francisco, was shipped away in large quantities, much of it to China.

By end of the 1870s the Selby Company was drawing from every important lead and silver locality on the Pacific Coast, including distant mining regions of New Mexico, Arizona, Mexico, and along the Colorado River, whence the ore and bullion was brought by sea via the Gulf of California. In 1879 the company expanded its operations by purchasing the Pacific Refining and Bullion Exchange. This business, which Mortimer Belshaw had an interest in, was an outgrowth of the refining and assaying business established in 1851 by Kellogg, Heuston & Company; and continued by the San Francisco Assaying and Refining Works. This acquisition gave Selby another important department. Gold and silver could be deposited for refining, the charges for this work being the same as at the Mint. Returns for gold bullion were made within 24 hours; 10 to 14 days on silver bullion. Payment of silver was made in fine bars which could be sold on the open market. This was the only private refinery on the Pacific Coast where miners and dealers in gold bullion throughout the country could make shipment by Wells, Fargo & Company, and receive their returns within 24 hours after being received by Selby. Payment was made in gold coin or by check, as shippers directed.

Eventually the Selby Company outgrew their North Beach location, and in the mid-1880s all its furnaces and old equipment were moved to a 43 acre site at Vallejo Junction in Contra Costa County.[28] There the company went on until purchased by the Guggenheim smelting trust about 1910. Ten years later suits over pollution shut the company down. More will be said of Selby in subsequent chapters. Now back to Inyo County.

When Cerro Gordo begun to wane in the mid-1870s Belshaw, Judson, and Beaudry bailed out. Belshaw left Cerro Gordo in 1877 and settled with his wife and son at Antioch, California. There, he founded a water works, a store and a bank. He was not finished with mining however. With friend Egbert Judson, Belshaw indulged in gold mining in Calaveras and Amador counties. In 1881 he and Judson obtained at sheriff's sale the Empire coal mine at Mount Diablo, which they revitalized by building a narrow-gauge railroad to deliver coal to Antioch. An unfortunate incident ended the relationship between Belshaw and Judson.

On July 9, 1892, an explosion wracked the Giant Powder Works at Fleming Point. Being a short distance away from the plant when the blast occurred, Judson jumped in his buggy to drive to the scene. The horse ran away with him, going toward the fire. Just then a second explosion occurred, and Judson was thrown from the buggy, striking his head. He lingered for six months before dying of heart failure during the week of January 14, 1893. Egbert Judson was 81 years old and unmarried. His estate, valued between $2 and $3 million, was divided between four nephews and nieces.

Belshaw's end also came from the effects of being thrown from a buggy. In April 1898, at age 68, the portly man was pitched from his buggy on a San Francisco street and died shortly after the accident. His wife Jenny returned to San Francisco, where she died in 1900.[29]

After the Cerro Gordo mines shut down Beaudry settled in Los Angeles and engaged in real estate. Extremely wealthy, Beaudry was interested in many public enterprises. An owner of extensive property along Temple Street, he was one of the developers of the Temple Street cable railroad. He married in Canada in 1875, and after residing in Los Angeles for many years, returned to Montreal in May 1887. He died there in early March 1888 of inflammatory rheumatism, leaving an estate valued at $1,500,000 to his wife and five children.[30]

What happened to Remi Nadeau? In addition to solving Cerro Gordo's shipping problems, Nadeau's operation made him the most successful teamster in California, and one of the state's wealthiest men. In 1879 Nadeau abandoned the Inyo trade and moved a large portion of his Cerro Gordo Freighting Company to southeastern Arizona. By early 1881 he had the choice contract in the Tombstone district, hauling ore for E. B. Gage's Grand Central Mining Company. A year later, in February 1882, Nadeau sold out his Arizona freighting business to Jimmy Carr and retired to Los Angeles where he not only had the largest vineyards in town, but a four-story hotel, the best in the region. Nadeau died in Los Angeles on January 15, 1887.

* * *

1. The accepted version of the discovery is that in March 1865 Pablo Flores and two other Mexicans started from Austin, Nevada, on a prospecting trip. They traveled southwest, investigating mountain ranges as they went. They did not find anything until they reached the area about Lone Pine. On a high peak called Cerro Gordo, on the westerly slope of the Coso Mountains, they discovered a number of argentiferous lead deposits.

2. Cerro Gordo is nearly 28 miles due east of Mount Whitney, 260 miles east and about 100 south of San Francisco, and 27 miles east of Visalia. The locality is 8,200 feet above sea level. Although their strike was at an elevation between 6,000 and 7,000 feet, a thousand feet

above the adjacent valley, the Mexicans deemed the veins rich enough to warrant exploration. Supplies were needed to carry that on, and it was agreed that Flores would remain at the site of the strike while his two companions went to Virginia City to secure grub and other essentials.

As the weeks passed, Pablo's companions did not return and he feared they had been killed by Indians. Nearly out of provisions, he set out to look for them. Making his way to Virginia City, Flores learned nothing about the fate of his friends. No one in the mining town had seen or heard of them and Flores concluded Indians had probably killed his companions. He then related details of his find in the mountains to the southwest, and induced a number of Mexicans to return with him to Cerro Gordo. *Mining and Scientific Press,* April 11, 1874.

3. Chaput, "Cerro Gordo," *Silver in the Golden State*. Oakland, Oakland Museum, 1986 p. 75.

4. There is considerable literature on the Catalonian furnace, a high bloomary forge of great antiquity, having been used in northern Europe since pagan times. Introduced into the New World by the Spaniards, the furnace underwent little modification during the last three centuries. Such furnaces were in use as late as 1881 at Willow Springs, four miles from Mineral Hill in Pinal County, Arizona. You may want to consult: *Mining and Scientific Press,* February 11, May 27, June 10, 1882; July 20, 1889; and February 23 and March 2, 1889; H. B. Pulsifer, "The Metallurgy of Lead," *The Salt Lake Mining Review,* August 15, 1912, p. 16. See also A. Gmehlino, "Smelting in Adobe Furnaces and Refining in Mexican Vaso," *Mining and Scientific Press,* January 3, 1885; and Henry F. Collins, "On Adobe and other Makeshift Furnaces," *Engineering and Mining Journal,* July 11, 1903, pp. 51-53.

5. The *vaso* is well described by W. L. Austin in "Mexican Cupellation-Hearth," *Engineering and Mining Journal,* February 21, 1885. That article is a condensation of a larger paper presented at the Chicago meeting of the American Institute of Mining Engineers, May 1884. See also Henry F. Collins, "On Adobe and other Makeshift Furnaces," *Engineering and Mining Journal,* July 11, 1903; and W. M. Brodie, "Metallurgy of Native-Silver Ores of Southwestern Chihuahua," *ibid.*, Vol. 101, No. 7, January 12, 1916, pp. 297-298.

6. *Mining and Scientific Press,* December 14, 1867.

7. James Delavan returned to his mining enterprises at Virginia City. In the late 1880s he became a special correspondent for the *Financial and Mining Record* of New York, contributing many articles about the Comstock. On August 27, 1891, he fell at his Virginia City home and fractured his left hip. Despite the "best medical skill or human sympathy," Delavan's condition worsened, and he died on October 22, 1891. He was survived by a son, a Virginia City businessman, and a married daughter. James's wife died about 1884. *Territorial Enterprise,* October 23, 1891.

8. Remi Nadeau, *The Silver Seekers.* Santa Barbara, Crest Publishers, 1999, pp. 14-15.

9. *Engineering and Mining Journal,* January 14, 1893, p. 36; Remi Nadeau, *The Silver Seekers,* p. 40.

10. *Mining and Scientific Press,* August 22, 1868; September 26, October 3, 1868; March 4, 1871. A. B. Elder discussed making the water tuyere in *ibid.*, September 10, 1898, p. 257. He stated that a water jacket, three feet around, came a year later. "It was gotten up by . . . Ira P. Rankin of the . . . Pacific Iron Works."

11. *Ibid.*, December 5, 1868.

12. *Ibid.*, January 23, 1869.

13. Contract: Perre Desormeaux and Victor Beaudry, January 30, 1868. Inyo County Recorder's Office, Agreements Book A, p. 30.

14. *Mining and Scientific Press* January 23, 1869.

15. The *modus operandi* of Desmoreaux's blast furnace was described by correspondent "Crowquill" in the *Mining and Scientific Press* of June 18, 1870.

16. Agreement, Beaudry and Desormeaux, October 10, 1868, Inyo County Recorder's Office, Agreements Book A, p. 50; *Mining and Scientific Press,* August 22, 1868.

17. *Ibid.*, January 2, February 20, April 10, June 26, 1869.

18. The Eclipse Mining Company, which had offices in London, also owned mines in Brazil, Australia, and England. *Ibid.*, July 23, 1870; April 22, 1871.

19. Chaput, "Cerro Gordo," *Silver in California,* pp. 83-85.

20. Francis Drake, "Early Smelting at Cerro Gordo," in *Mining and Scientific Press,* May 21, 1910.

21. *Mining and Scientific Press,* November 21, 1874.
22. *San Francisco Chronicle,* June 10, 1875.
23. A description of these works is found in the *Mining and Scientific Press,* February 23, 1867.
24. *Ibid.,* May 4, 1867.
25. *Ibid.,* July 16, 1870; December 26, 1874.
26. *San Francisco Chronicle,* June 10, 1875.
27. One observer writing in the *Mining and Scientific Press* of September 16, 1876, described Selby's refining procedure: "The ore, landed at the wharf, is brought by railroad into the works, where it is crushed, sampled and prepared for calcining; thence it goes to the blast furnaces to be smelted; thence to the refining furnaces, where it is cleared of base matter; thence to the desilvering furnaces which separate the silver from the lead. Here it "splits" and takes two directions — the lead going to the refining furnace again, where it is converted into market lead, is stamped with the proprietor's name, and is piled away ready for shipment. The residue from the desilvering furnace goes back to the smelting and thence to the cupel furnace, where the small percentage of lead still remaining is extracted, leaving the silver pure or nearly so. This is melted again in crucibles to still further refine it; for although it comes from the cupel furnace 990-1000 fine, which is suitable for the purposes of the Mint, it is not fine enough for shipment to China, where the standard is 996-1000. The perfection to which this art may be brought is shown in one lot of 10,000 ounces, which assayed nine hundred and niney-nine and three-tenths one thusandths fine."
28. More information on the new plant is found in the *Mining and Scientific Press,* April 17, 1886.
29. Remi Nadeau, *The Silver Seekers,* pp. 14-15, p. 40; *Engineering and Mining Journal,* January 14, 1893, p. 36.
30. *Los Angeles Evening Express,* March 8, 1888.

Eureka, Nevada, ca. 1870. Courtesy National Archives, Washington, D.C.

Chapter Three

FREIBERGERS AND THEIR FURNACES: WHITE PINE AND EUREKA, NEVADA

The next plunge into smelting occurred in eastern Nevada and coincided with the Cerro Gordo boom. That story began in autumn 1865 when a party of prospectors from Austin were attracted to the region east of the Diamond Range by a view of mountains covered with white pine timber. Instead of harvesting lumber they discovered outcrops of silver, lead, and copper. A mining district, called White Pine, was organized on October 10, 1865.

In 1866 Thomas J. Murphy and J. W. Crawford went to Philadelphia with specimens of silver ore from claims in the White Pine District. Wealthy Pennsylvanians thought the ore rich enough to gamble on the claims, and the Monte Cristo Mining Company was organized. A superintendent was sent out in 1867, and a mill was put up to work the ore.

What really got the ball rolling was the appearance of an Indian in autumn of 1867. For some trifling favor bestowed by A. J. Leathers, the blacksmith of the original company, the Paiute handed over a piece of ore which upon melting produced a button of silver. The Indian was induced to reveal where he found the specimen, which proved to be a prominence near Hamilton, later called Hidden Treasure Hill. The Hidden Treasure claim was located January 3, 1868, by Leathers, Murphy and Edward Marchand, and sold to G. E. Roberts & Company in January 1869 for $200,000. Soon after the discovery of the Hidden Treasure, T. E. Eberhardt, of Austin, discovered the famous chloride deposit on Treasure Hill known as the Eberhardt Mine.

The Eberhardt Mine was disposed of in 1868 to a company which took several tons of the ore to Austin for reduction, where it was found to yield from $450 to $27,000 to the ton. Ore yielding

$3,000 to the ton was constantly taken from the Eberhardt, Keystone, and Blue Belle mines, which were banked up for smelting, should furnaces be erected. On September 25, 1868, the Defiance Mine produced in one day ore that would yield $40,000 worth of bullion; and $75,000 worth of ore was still in sight. This prodigious wealth created the greatest excitement since discovery of the Comstock. Thousands of men hastened to White Pine, and the prospector's pick was heard in all directions. Within weeks the canyons about Treasure Hill were adorned with miners' cabins. By fall 1869 there were 165 stamps pounding away on "free-milling" ore, and 30 more were being erected.[1]

In their quest for silver, prospectors bypassed a tremendous amount of "base metal" ore (i.e. sulphides, oxides or carbonates of copper, iron, lead, zinc or antimony). As these ores could not be worked by amalgamation, they were deemed valueless; the deposits laying undisturbed until 1868-69 when the argentiferous galena lodes of the Eberhardt and Hidden Treasure mines were encountered.

Attention was then focused on White Pine Mountain. Every stain of green, brown or blue, whether on quartz, limestone, or dolomite, was located; hundreds of veins discovered. The principal claims consisted of carbonate of lead, assaying 50 to 55 percent lead, and from $15 to $29 in silver. Ores mixed with copper went as high as $150 per ton. Because of the lead content of its deposits the mountain was renamed the Base Metal Range.

Base metal deposits had advantages over those of "free milling" ore. The yield of ore was more regular, the claims more easily worked, and the ore more saleable from the dump. The reason for this being that the value and character of carbonate ore and of argentiferous galena could be seen at a glance, and it could be cheaply reduced by smelting. On the other hand chloride ore had to be milled to determine its value. Even with a great difference in silver yield, base metal ores were more profitable.[2] In short, this ore presented an opportunity for poor men to own and work claims, and just like Cerro Gordo, poor men were the first to erect smelters at White Pine.

With a view of avoiding the expense of blowers and engines, many White Pine miners built reverberatory furnaces, which could be fueled with any kind of wood. Though many of these smelters were built with chimneys to increase draft, reverberatory furnaces lost one-half of their heat. Other Base Range miners partnered in the erection of draft furnaces. Constructed at trifling expense, these small furnaces were identical to Catalonian furnaces, except they had no mechanism to create a blast, not even bellows. In the end blast furnaces prevailed, for the simple reason that they were economical to operate, even in a region where fuel, namely charcoal, was expensive.[3]

When it was seen that argentiferous lead ores could be successfully worked, entrepreneurs moved in. Several upright cupolas, built of common country rock lined with sandstone, were erected at Swansea in Sawmill Canyon. The blast was obtained by a horse-powered McKenzie or Sturtevant fan. Several tons of ore were smelted, then the fan burst and the horses ran away. A second attempt to run by steam power failed; the roof caught fire and the concern burned down.

By then there were others in the act. The San Francisco Smelting Company, owned by Welshman Richard P. Jones, erected works at Swansea, a short distance above Shermantown, as did J. J. Bassett. W. H. Pritchard & Company, a San Francisco-based firm, built six blast furnaces in Mohawk Canyon on the east slope of the Base Range. A 20-horsepower engine drove the blast. By October 1869 smelting furnaces were "multiplying." There was Tucker & O'Hara's furnace; Sanford Hall's furnace at Shermantown, and the Shields Smelting Works two and a half miles west from Hamilton. Also at Hamilton were three furnaces put up by J. R. Partridge and Capt. Magary. Jones, Hall & Bradley had another smelter at Shermantown, as did Mesen & Higgins. An assayer named Van Wyck built still another smelter within the town of Hamilton, and between Swansea and Shermantown was Phillips' blast furnace. At least 40 furnaces of various shapes and styles were in operation within a year.[4]

Built by men versed in mining and not the science of metallurgy, most of these furnaces failed for a number of reasons. Excessive blast overheated the charge and burned out the lining around and above the tuyeres. Hearth failures, however, were mostly due not to the fault of refractory material but to the incompetency of the proprietors. In most cases furnace hearths were not properly cured.

This part of the furnace was made of clay and charcoal rammed into solid brickwork and then cut out to the desired shape. If not allowed to gradually dry, and then heated up gently for a day or two to drive out the moisture, or if the furnace was heated up too soon,

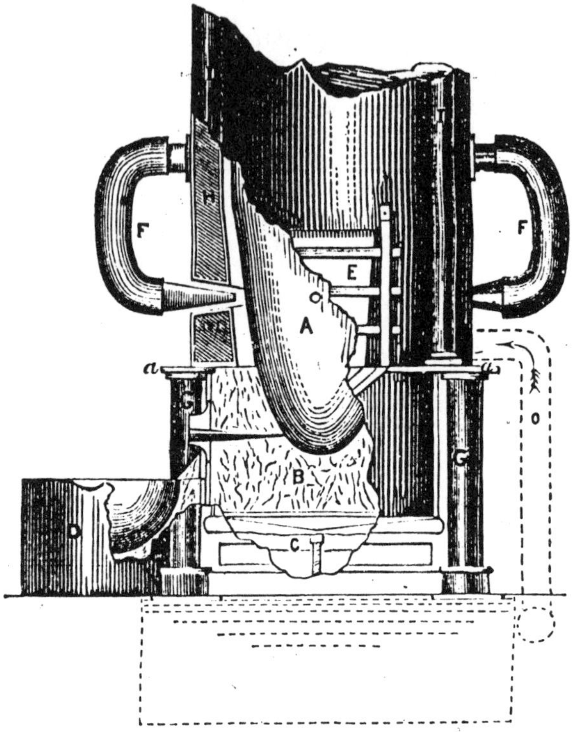

Mosheimer's furnace with movable hearth as depicted in the *Mining and Scientific Press* of June 11, 1870. The upper part of the furnace was constructed of masonry. The part subject to the greatest heat consisted of a wrought-iron shell, H, which acted as an air-chamber protecting the lining of the furnace. Air produced by a blower passed through the pipe, *o*, into this shell, keeping it cool, and thence through the tuyeres, F, into the furnace. This shell rested on an iron plate, *a*, supported by four posts, G, one at each corner. The upper part of the furnace was also supported independently on iron posts, I. The shaft was lined inside with three inches of fire clay to protect the iron. Furnace bottom or hearth, secured in place by screws, was easily moved in and out on wheels. It was made of strong boiler iron, either round or square, filled with sandstone from White Pine, pounded down and cut into shape. An extra hearth was kept constantly on hand to replace those burned out. Cost of iron work in San Francisco for such a furnace was $700, including freight.

the escaping moisture caused the metal to boil, cracking the mass. When that happened metal impregnated surrounding bricks or stonework. In some cases, furnace bottoms were found diffused with several tons of metal, so solidly cemented by lead that it took days to cut them out. Such catastrophes created expensive repairs, delayed production, and resulted in loss of metal which could not be re-smelted. In one case, an improperly cured hearth of a White Pine smelter exploded, knocking out its front and rear walls.[5]

There were two notable successes: the Alsop Smelting Company at Shermantown, organized by Capt. Rawlings, C. Carpenter, F. S. Luther, and Thornburg, and the White Pine Smelting Company at Swansea. Both were San Francisco companies.[6] Because they dominated the smelting scene for many months, we'll look at them in detail beginning with the White Pine Company.

The White Pine Smelting Company was organized in San Francisco by metallurgist Joseph Mosheimer. An Austrian, born about 1823, Mosheimer attended the School of Mines at Freiberg, Saxony. According to the *Mining and Scientfic Press* of December 12, 1868, Joseph "engaged in the special field of metallurgy on this coast even earlier than the discovery of silver in Washoe. Mr. Mosheimer is, like his works, an epitome of all that has taken place." That is certainly true, for Joseph arrived in San Francisco in 1851, probably as an assistant to Freiberg graduate, Guido Kustel, the latter six years his senior. They opened the first assay and metallurgical business in the Bay City. In September 1861 they refined the first ore from the Comstock, the bullion from which was put on display in the banking house of Alsop & Company. This exhibit infected the public mind with the idea of silver mining in Nevada. Kustel and Mosheimer joined the rush to Washoe, both being employed building mills along the Carson River.

After the amalgamation process became understood, Joseph devoted himself to the treatment of refractory ores and to smelting. In 1868 Mosheimer occupied a large building at the rear of No. 2005 Powell Street, near North Beach in San Francisco. There, he designed and built furnaces, both for roasting and smelting, and instructed students in the art of metallurgy at a tuition ranging from $150 to $250 depending on length of instruction. With impeccable credentials, Mosheimer was besieged by mine developers seeking assessments of their properties and advice on how to process ores, among them Mortimer Belshaw, Abner Elder, and Victor Beaudry.

So it was natural that Joseph Mosheimer would be drawn into the White Pine bonanza, either as a consultant, investor, or both. Sensing opportunity in the vast deposits of base metal, he organized a company to process argentiferous galena that was beginning to flow from claims at White Pine. Designing a smelting furnace that was in some respects quite revolutionary, he assembled the components in San Francisco, and in summer of 1869 headed for White Pine.

As superintendent of the White Pine Company, Mosheimer in fall of 1869 erected a trapezoidal shaped shaft furnace, 24 inches to a side and 12 feet high. The lining was made of a fireproof sandstone found close by. The blast was supplied by a four-cylinder steam engine driving a revolutionary device, a Roots blower.

By choosing the Roots blower, Mosheimer demonstrated a knowledge of the latest advancements in metallurgy. The world's first successful positive blast rotary blower, this device had been on the market only a few years. Invented in 1860 by Francis M. Roots[7] and his brother P. H. Roots, the blower consisted of two revolving vanes geared together so as to keep up constant contact with each other, and with the internal periphery of a pair of half cylinders. The Roots brothers built two of these blowers of sufficient size for large foundries. Tested for two years at the Cincinnati, Ohio, works of Miles Greenwood, and in the stove factory of G. W. Ball at Covington, Kentucky, the blowers gave a more regular, steady, and powerful blast than the ancient fan type. Doubling the melting rate of pig iron, the invention was deemed a success, and the Roots brothers purchased a small machine shop and foundry in Connersville, Indiana, and began manufacturing blowers in 1863.

The smelting industry quickly recognized the merit of the Roots blower. Sales took off, and from 1863 until 1874 the company was frequently enlarged. The blowers were sold throughout the United States and Canada, and sent to every part of the world where machiney was used. The Roots blower ultimately replaced the McKenzie and Sturtevant fans in the American smelting industry, and became as much an article of commerce in the machinery world as the portable engine or the screw cutting lathe. Today Roots blowers supercharge diesel trucks and race cars. Now back to White Pine and Joseph Mosheimer.

Mosheimer built a second furnace, double the size of the first, of sandstone blocks 12 inches thick. The structure did not need an out-

SECTION OF BLOWER.

Above: section of the Roots blower; below: a Roots blower of the 1880s. (From *Scientific American*, April 3, 1886)

side shell or wall to keep it together, for it was well bound with iron. Built in a "workmanlike manner," the furnaces cost $20,000.

The best quality charcoal, made from mountain mahogany and nut pine, was stockpiled, as was 400 tons of ore purchased from nearby mines. In late October 1869 the smelters were blown-in under direction of a young German metallurgist named Otto Hahn, of whom more will be said. The first campaign consisted of 15 tons of lead carbonate from the Miser's Dream Mine. The yield was 55 percent lead and about $30 in silver. A small addition of clay or loam answered as flux, and about 15 to 20 bushels of charcoal were used to the ton of ore. The cost of charcoal was 30 cents per bushel delivered at the smelter.

Running two furnaces, one with a daily capacity of 24 tons, the other of eight tons, the White Pine Smelting Works produced on average four tons of bullion per day, worth $90 in lead and $50 silver to the ton. Thirty-five tons of bullion was poured in the first week of operation. Mosheimer's smelter was heralded an immediate success, to the benefit of not only the mine it was associated with, but to many independent miners as well.[8]

Midwest and eastern refiners moved to capitalize on White Pine's bullion output. Because the cost of transportation was as low to Chicago as to San Francisco, Chicago refiners such as the Chicago Silver Smelting & Refining Company, the Swansea Silver Smelting & Refining Works, and S. P. Lunt Company, contemplated erecting separating works at Hamilton.[9] In the end, however, the Balbach Refinery at Newark, New Jersey, cornered the White Pine bullion output, at least for the moment. Arrangements were made with the Bank of California at Hamilton to make advances upon crude bullion shipped to Newark.

At 1869 Balbach was the largest refinery in the United States, having been around since 1849. Located on the Passaic River in the city of Newark, the company was the creation of Edward Balbach, who was born in Carlsruhe, Baden, Germany, March 19, 1804. Balbach studied chemistry and engaged in the refining of metals in his native city. Government restrictions of the precious metals trade, however, curtailed the growth of his business, and in 1848 Edward visited the United States with the view of setting up business. He selected Newark, scarcely 35,000 inhabitants, but a center for the manufacture of jewelry. Moving to the United States in 1850, Balbach laid the foundation for his company.

With skill and fair business methods Balbach secured the confidence of local jewelers, refining their scraps of precious metals. It did not take long for his reputation to extend beyond local limits to include the principal East Coast jewelry manufacturers. He was soon receiving small shipments of ore. In 1859 Balbach took on the refining of ore from a lead mine in Orange County, New York, and a year later he was receiving ore from a Pennsylvania lead mine. A few years more and the fame of his Newark works reached to remote parts of the United States.

In 1866 large shipments of base bullion from Nevada furnaces found their way east and were treated by the cupelling method, a slow and expensive procedure to separate gold and silver from lead. The desire for a more practical method caused Edward Balbach, Jr., who had joined his father's business, to seek profitable ways to refine high-grade bullion. He experimented with the European method utilizing zinc to separate precious metals, but encountered difficulties in the distillation of metallic zinc when the lead destroyed the retorts. He remedied that by inventing a black-lead retort. Patented on May 21, 1867, this apparatus and its methodology, known as the Balbach desilverization zinc process, was adopted by American and European smelting establishments. This process assured the continued success of the Balbach refinery. Building after building, furnace after furnace, for the best treatment of all classes of argentiferous and auriferous material, were added to keep pace with constantly increasing demands. By 1870 Ed Bulbach, Sr., was recognized as the "Nestor of the gold and silver-lead smelting industry," the firm dominating the bullion trade of the Far West.[10] Its position, however, would be contested by the Selby refinery several years later. More of that later.

Back at White Pine, the situation at Mosheimer's smelter was critical. Smelting ores were "eminently basic"; poor in silver, abundant in lead carbonate. Easily treatable, it was thought. Owing to an absence of quartz in the ore, and the difficulty of obtaining siliceous ore for fluxing purposes, their reduction gave rise to formation of "sows innumerable," iron concretions which hindered smelting and often shut down furnaces. This was remedied by O. H. Hahn, who searching for a substitute, hit upon the use of clay slates, abundant in the neighborhood, which resulted in the first attempt to make aluminate instead of silicate slags. The experiment was successful.

Behavior of fuel was also a great annoyance. At White Pine most charcoal was burned from mountain mahogany, a firm compact wood. But the process of carbonization burned the wood out, producing a soft charcoal. When put into the furnace and the charge thrown upon it, it became crushed to an almost impalpable powder, which blocked the blast. In some instances the fuel actually put the fire out. Nevertheless, these works produced in the course of a few months more than 10,000 tons of bullion, resulting in a gross financial loss of $30,000, which caused a temporary shut-down of operations, and the resignation of Joseph Mosheimer as superintendent. His position was filled by 31-year-old Winfield Scott Keyes.

Born in Brooklyn, New York, in 1839, Keyes graduated from Yale College in 1860, and three years later from the School of Mines at Freiberg, Saxony — educational attainments that earned him the sobriquet of professor. It is doubtful he ever taught, for he was more involved in metallurgical work. For several years he worked in San Francisco, which put him in contact with Joseph Mosheimer and others involved with the White Pine Smelting Company.[11]

Thus Keyes was the man chosen to iron out the company's financial woes. Assuming the superintendency, Keyes fixed the price of ore at the dump at between $10 and $20 per ton, and raised the rate for smelting from $40 to as high as $75 a ton. The latter price was so "enormously and disportionately high as to prohibit miners from having it worked in that manner," which led to shutting down of furnaces, and abandonment of smelting for a while.

The smelting void was momentarily filled by the Alsop Smelting Company, located at Shermantown a mile farther from the Base Range than the White Pine Works. Organized in early 1870 by Rawlings, Carpenter, Luther, and Thornburg, this custom smelter consisted of a cupola furnace of 24 tons daily capacity similar in design to that erected by Mosheimer. To reduce copper matte produced from the large cupola furnace, the firm installed a Haskell water-jacket furnace.

Commencing operation in April on ore purchased from the Alta claim, the Alsop Company made a 34 day run, turning out 157,000 pounds of bullion. A second run produced 72,000 pounds of lead pigs. In two months it cleared all debts incurred for construction and repairs, and made a handsome profit. Because it paid better prices for ore than the White Pine Company, Alsop was favored by miners, a situation Keyes tried to counter.

Stating that "business was unprofitable," Keyes fixed the price of carbonate ore on the dump at $7.50 or less per ton. On top of that the smelting company required the miner to delivery the ore. The effect, according to a *Mining and Scientific Press* correspondent, was devastating:

> A man with a family dependent upon him for support, loaded a wagon with a couple of tons of ore, of a quality that had previously commanded from 20 to 30 dollars per ton, and took it to Hamilton expecting that the new company would pay something near its value, when offered to them. What was the poor man's astonishment when told that he could have $8 a ton for his ore! He shoveled out his load and took his departure, with a vow that this company should have no more at any price.

If that was not enough, Keyes endeavored to induce managers of the Alsop works to enter into a "combination" to keep down smelting prices. While he was doing that, the White Pine Company was building a still larger furnace to compete with the Alsop works. The net result was that "no man can live by his labor in the Base Range." By October 1, 1870, "there were not 25 men to work . . . their claims" in the district. There were less on the spurs of Treasure Hill, and about 20 men remained on White Pine Mountain.

Although the White Pine District was in a dismal strait some still saw profits in smelting the region's ores. In September 1870 Joel Matteson, ex-governor of Illinois (1853-57) and president of the Chicago and Alton Railroad, in association with other capitalists, erected the district's largest smelting operation. Three large cupolas were built at Hamilton, with forced blast by means of a 40-horse-power engine. Five roasting kilns were also erected to prepare ores for the furnaces. Separation of precious metal would also be performed. Boilers, engine, blower, and refining furnaces, were all of Chicago manufacture. Enclosed in a substantial stone building, the works occupied a block of ground. (In all, four blocks belonged to the Company.) A "never-failing" spring upon the property was converted into a reservoir to supply the building and boilers. Total cost of the smelting and refining works was in the neighborhood of $50,000. Having a large amount of capital, the company was determined to "push things."

In "pushing things," it actually pushed the miners to the wall. The managers insisted miners deliver ore to the smelter. They would then assay it, affix their own valuation of the various lots. If the miner was not satisfied, he had the alternative to cart it away at his own expense.[12]

In summer 1870 only nine smelting works were still operating at White Pine. Ninety days later the White Pine Smelting Company failed; its officers settling their obligations at from 30 to 50 cents on the dollar. An effort was made to reorganize the company in New York, but that failed. On January 3, 1871, the White Pine Smelting Works was sold at sheriff's auction. Of 20 people attending the sale, only one made a bid: $2,000 for personal property, $500 for real estate. The offer was accepted.[13]

In contrast, before ceasing operations the Alsop Company paid every debt to operatives and miners, and at the highest prices. The Rathburn Brothers stopped work with their furnace, sent nearly all their bullion away, and then followed themselves, leaving the worn out furnace to settle all demands against them.[14] Several other furnace operators did the same thing, only with less loss to the mining community. By end of September 1870 not a furnace was operating in the district, and the Base Metal range was "almost wholly deserted." The cause of the stoppage was attributed to "a lack of capital to carry on the business even long enough to allow time for returns from bullion to arrive."

"It was not that 'the bottom has fallen out of White Pine,' as some have asserted, nor even out of the furnaces," wrote a correspondent in the *Mining and Scientific Press*, "but that the pseudo-capitalists who have tried smelting here as a business, have had neither bottom nor integrity." Base metal smelting was not dead in Nevada, however.[15]

* * *

The next plunge into base metal smelting was at Eureka in Lander County, about 40 miles west of Hamilton and 65 miles east of Austin. In May 1864 silver-bearing veins were found near the overland stage station of Diamond Springs, and the Diamond District was formed. The discovery which named Eureka was made by a prospecting party from Austin composed of W. O. Arnold, W. R. Tannehill, G. T. Tannehill, J. W. Stotts, and Moses Wilson. On September 19, 1864, these men found a black, heavy ore differing from silver chloride so abundant at Austin. They experimented with the rock, placing it in a camp fire; it melted. When they learned the metallic globules were rich in silver, they supposedly exclaimed, "Eureka." A good story, probably without foundation.

A drawing of the first successful smelting furnace at Eureka, Nevada, erected in 1869 by G. C. Robbins. Initially a draft furnace, it was rebuilt as a blast furnace. (*Engineering and Mining Journal*, October 26, 1912)

Claims were filed and a mining district, first called Napias, a Shoshone word meaning silver, was organized in New York and Secret Canyons, with G. T. Tannehill as recorder. The district's name was later changed to Eureka. Moses Wilson built a smelter but failed miserably to reduce his ore. Nevertheless, prospecting continued for several years thereafter, and a modest number of claims were filed. Although good milling rock was located, much of the ores were sulphates, antimoniates, and carbonates of lead. Carrying from $20 to $200 in silver per ton, the ore was valuable but its high lead content rendered it unworkable by amalgamation.

Not knowing how to develop their properties, and lacking capital to experiment with ore reduction, the locators sold their "lead mines" to New York capitalists who expended a large sum of money. Lacking proper knowledge of reduction techniques, they too were unsuccessful.

An entirely different class of ore existed on Mineral Hill, two and a half miles southwest of where the town of Eureka would develop. Discovered in 1865, this ore zone was extensive, actually two parallel mineral-bearing belts running for miles; deposits of greater or less magnitude cropping out in limestone above a quartzite horizon. Carrying a high percentage of base metals, this ore was much richer in silver than at White Pine, in some cases running as high as $500 to $600 per ton.

In the meantime, a Paiute Indian showed Alonzo Monroe, M. G. Clough and Owen Farrell some black rock which bore indications of silver. For ten dollars the Indian guided the prospectors to a spot on a northwestern spur of Prospect Mountain, about two and a half miles west of Eureka, subsequently called Ruby Hill. They located the whole hill, the most important claims being the Buckeye and Champion, the latter having an immense orebody. In spring of 1868 Morris, Monroe and others employed Charles A. Stetefeldt, an Austin assayer, to work what was termed "smelting grade ore."[16]

At the time Stetefeldt was probably the foremost metallurgist working in the Far West. Born September 28, 1838, in the village of Holzhausen, in the duchy of Gotha, Germany, Charles was the only son of a Lutheran pastor. Young Stetefeldt began his studies in physical science because of an aversion to Hebrew, which his father wanted him to study in preparation for a career in theology. So far as Charles was concern, geology far outweighed religion, and in 1858 he matriculated at the University of Gottingen; two years later he entered the School of Mines at Clausthal, and passed the "engineer's examination" in 1862, receiving in all branches of mining and metallurgy the highest degree. He was at once commissioned to investigate the Pattinson process, and the causes of "over-poling" of copper at the government works in the Harz. After visiting Freiberg, Mansfield, Stolberg, and other metallurgical centers, he took on the technical management, for a brief period, of a small copper-plant in Bohemia.

Emigrating to the United States in 1865, Charles was engaged as an assistant by Charles Joy, professor of chemistry at Columbia College in New York City. The following year Stetefeldt joined the New York City firm of Justus Adelberg & Rossiter W. Raymond, mineralogists and consulting metallurgists. The timing of that association could not have been better. American mining was just getting underway and experts were scarce. Consequently, the firm of Adelberg & Raymond was overwhelmed with work, requiring the assistance of many young men, mostly graduates of German schools. Thus Stetefeldt was thrown into contact with Anton Eilers, Otto H. Hahn, and Albert Arents, all of whom became recognized authorities in modern metallurgy. Of all of these young men, according to Raymond, "Stetefeldt was at that time the most thoroughly and widely accomplished. . . . He possessed a knowledge of mathematics and chemistry much beyond the usual equipment of a mining engineer or metallurgist. . . ."

In 1865 Stetefeldt and John Henry Boalt, a Freiberg classmate, set up branch offices of Adelberg & Raymond at San Francisco and Austin. Obsessed with desulphurizing pyritic ores, Stetefeldt used his time in Nevada to experiment using a modified Gerstenhofer's shelf-furnace. He subjected Reese River silver ores to the Gerstenhofer roasting furnace. His tests were unsatisfactory, the ores caking and forming crusts upon furnace shelves. He eliminated the shelves, increased the height of the furnace. Through trail and error, Charles learned that silver ores, no matter in which combination the silver occurs, mixed with salt, were completely chloridized if they fall against a current of hot air rising in a shaft with no obstructions to check the fall of ore particles. The result was the Stetefeldt furnace, invented at Austin in 1867. Thus Charles A. Stetefeldt came to the Eureka District with impressive credentials.[17]

Stetefeldt erected a one-tuyere smelter and made three attempts to reduce ore from the Buckeye and Champion claims; all failed. The failures were attributed to the "quartzy" nature of the ore, and its limited extent. More likely the failures were due to insufficiency of blast, poor furnace construction, and employee incompetency. Stetefeldt returned to Austin in disgust, and departed for Europe a year later.[18] His furnace, the first in the district, was placed in the hands of Major W. W. McCoy,[19] who hired a couple of Welshmen, Richard P. Jones and John J. Williams, Jr. Both were expert smeltermen.

Born at Swansea, Wales, in 1845, John J. Williams, Jr., received his early metallurgical training in the Vivian Smelting Works. He also had the advantage of instruction from his father, who was an authority in mining and reduction of metals. John Williams, Sr., a veteran of the Vivian works, introduced the water-jacket furnace to

Drontheim, Norway, in 1852. Six years latter he brought his family to the United States to engage in mining and smelting, first in Connecticut, then in the Lake Superior copper district. In 1865 John took on the management of the Portage Lake Smelting Works at Houghton, Michigan, where he erected a number of water-jacket blast furnaces for experimental smelting of Calumet ores. The rush to promote the West's mineral wealth drew John and his sons, Lewis, John, and Ben, to the Great Basin in 1868.

Thus John Williams, Jr., was a journeyman smelter when he reached Nevada. How he met Jones is not precisely known; they may have had a long relationship, or it may have commenced at White Pine in 1869 at the San Francisco Smelting Company of which Jones was a principle. What is known is that Jones and Williams introduced to Eureka Pancake Mountain sandstone, a perfect material for the lining of furnaces.

Jones and Williams rebuilt Stetefeldt's old furnace, inserting two more tuyeres, and lined the smelter with Pancake Mountain sandstone. In July 1869 they made their first run on 60 tons of ore from the Champion, Buckeye, Grant, and Eureka claims. The run was successful and 20 tons of bullion was shipped to San Francisco, assuring the future prosperity of the district.

In late 1869 G. C. Robbins, representing the Buttercup Mining Company of New York, arrived in the district and purchased the Kentuck and Mountain Boy claims. Robbins hired W. T. E. Pritchard, a son of a Welshman and a Mexican woman said to be "racially a metallurgical marvel," to erect a draft furnace. Although the crew of Mexicans and Indians celebrated feast days with great vigor, the furnace was finished in December, and ore from the Mountain Boy claim run through it. Both operations demonstrated smelting could, under skillful direction, profitably reduce the district's ore.[20]

Robbins' success drew the attention of entrepreneurs, the most prominent being I. C. Bateman and David E. Buel. The latter, a Texan said to have enormous feet, had been the Democratic candidate for governor of Nevada in 1864. Both men had figured in the creation of Austin, Buel being one of the promoters of the Austin townsite and a candidate for mayor. Both were also directors of the Humboldt and Colorado Railroad, a company formed at Austin in May 1868 to build a railroad from the Humboldt River to the Colorado.

Before the town of Eureka existed Bateman and Buel sensed money could be made in the district.[21] In November 1869 they ventured south from Austin to bond the Buckeye and Champion claims. Forming a San Francisco corporation, known locally as the Bateman Association, they leased McCoy's furnace, which they ran successfully until termination of their lease in May 1870. In the meantime, Buel and Bateman hired Richard P. Jones, who by then was considered a "first class workman who knew his business," to erect a two furnace smelter at the lower end of the canyon. The first furnace was blown-in on April 25, 1870, and the second was operating a few weeks later.

Both furnaces were built of stone found in the vicinity; their interior lining being Pancake Mountain sandstone. Furnace height from hearth to feed-hole was 16 feet; the stack 25 feet higher. The machinery and iron work consisting of a No. 7 Sturtevant fan, driven by a 38-horsepower engine and boiler, the latter being 15 feet long and 54 inches in diameter. Both engine and boiler were fabricated in San Francisco by the Miners Foundry and the Neptune Iron Works.

In July 1870 Buel and Bateman purchased the Buckeye for $100,000 and merged it with the Champion. The two claims, together with the smelter, formed the basis of the Eureka Consolidated Mining Company. It was the Buckeye and Champion, worked to a large extent as open quarries, that proved ore continued in depth. Two shafts, respectively 75 and 65 feet in depth, were sunk and the two connected by a level. On this level a chamber 16 feet square was cut out in the ore, which did not touch the limits of the deposit on either side.[22]

The Eureka Consolidated smelter commenced production in December 1869. The two furnaces, under supervision of Joseph Farren, each worked 13 tons of ore per day, producing about seven tons of bullion, assaying $140 to $180 to the ton in silver and $120 to $160 in gold. Such production caused a proliferation of smelting companies. The Philadelphia Company (Wallace and Bevan) erected two furnaces, their operation failed, however. The Robbins furnace, constructed of iron with fireproof brick lining, and converted to a blast furnace, commenced operation in mid-May 1870. This was followed by McCoy's furnace, managed by McGee and Carpenter, (formerly of the Alsop works at White Pine), and Wilson's furnace; McNevin's smelter was the sixth. Two more furnaces were constructed in August 1870. A month later there were ten furnaces in the district, and three more under construction.[23]

Smelters multipled over the the next several years. In 1871 a London-based corporation purchased the Richmond Mine, a series of patented and unpatented claims covering a large portion of Ruby Hill.[24] Forming the Richmond Consolidated Mining Company, the firm acquired the smelting works of Ogden, Dunne & Company. With English thoroughness, four furnaces and a refinery to separate precious metal from the lead, were added to the original works. By 1876 there were 19 furnaces in operation at Eureka, whose daily capacity varied from 40 to 60 tons each.

Running the same ore — cerussite or carbonate of lead, and some galena — these furnaces were similar in design. They were vertical blast or cupola furnaces with a square horizontal section, built of stone found locally, lined inside with porous sandstone from Pancake Mountain. Fan blowers furnished the blast, which was admitted through three tuyeres, one at the back and one on each side. The mouth or nozzle of the tuyeres were three inches in diameter, admitting a large amount of wind; in some cases too much, considering the easy fusibility of ore and the horizontal dimensions of the furnaces, which were about two feet square. Furnace shafts, from tuyeres to the charging hole, were from 12 to 16 feet high. The hearths about two feet below the tuyeres, were made of stone, the front part alone being formed of a composition. Tap-holes were placed at the bottom of the hearth.

About 24 pounds of charcoal were charged into the furnaces at one time, along with 40 to 45 pounds of ore, and 12 pounds of slag. Charging was done in an irregular manner, by means of shovels, without weighing or measuring. One furnace could smelt from eight to nine tons of ore in 24 hours. Three and a half tons of ore yielded about one ton of pig lead, or "bullion." Consumption of charcoal varied from 30 to 35 bushels to the ton of ore. The cost to smelt a ton of ore in 1870 was about $20.[25]

Tapping the furnace was a laborious procedure. The blast was stopped and the tap-hole in the bottom of the crucible opened, sometimes with great difficulty, when metal had cooled in it at a former tapping. The matte and slag was run out into the kettle, the hole was stopped again with clay or a mixture of clay and coal dust, called "stubbe." The blast was turned on again and smelting resumed. With cleaning of the crucible, and building up of forehearth, this part of the smelting procedure often took considerable time. This, of course, reduced the temperature in the furnace and wasted fuel.

Irregularities of furnace running were frequently traceable to the lengthy tapping procedure. Formation of "sows" occurred in nearly all cases during stoppages, when small doughy masses of reduced metallic iron stuck to the bottom of the crucible, which was no longer protected by a liquid mass. These grew rapidly, and despite being chiselled out at every tapping, iron concretions continually formed within the furnace.[26]

Bullion was run out at intervals, one tapping giving about 200 pounds of lead, which was run into moulds forming bars called pigs, weighing from 80 to 120 pounds apiece. When furnaces were properly managed loss of lead did not exceed 20 percent. Other factors could drastically increase the loss. Frequent use of crowbars about the hearth to break up sows could mix a great deal of lead into the slag. Large tuyeres bringing too much wind into the furnace diffused heat to upper parts of the furnace, instead of concentrating it in the smelting region just above the tuyeres. In such an atmosphere lead carbonate will melt at a distance of one or two feet below the charging hole, causing the lead to vaporize. This situation also lowered the hearth's temperature, causing the slag to stiffen and stick to the walls, requiring constant use of crowbars. And lastly, very long runs in furnaces in need of repairs contributed to loss of lead and silver.

The bullion was transported by teams 90 miles to Palisade, a station on the Central Pacific Railroad, and consigned to Newark, New Jersey, to be refined and separated. The cost for separating the lead and silver was $15 per ton; $7 extra was tacked on for extraction of gold. The Balbach works allowed $1.29 per ounce for the silver, and 6½ to 7 cents per pound for lead.[27]

The big hitch was that a continent separated Eureka from Newark, New Jersey. It took a long time for cash to reach the smelters and the mines. In the meantime, transportation, labor, and repairs costs ate up operating capital. In August 1870 a correspondent reported "money appears to be scarce here just now. I am told that the smelting works have much trouble in getting returns for the bullion they have shipped East. The camp has grown up on its own merits, and not a cent of capital has come in until of late."[28] Consequently, Eureka smelters looked to a closer source for processing their bullion. The Selby Refining Works of San Francisco gave them an alternative.

Smelting at Eureka in particular, and in the Great Basin generally, would be revolutionized by the appearance of metallurgists who received their early training in two German mining regions, in the Harz and at Freiberg. The former, sometimes spelled Hartz, was a mountainous strip in central Germany between the Elbe and Weser rivers. For hundreds of years argentiferous ores had been mined in the "Upper Harz," the area about Clausthal, Lautenthal, Altenau and Andrensberg. Early on a school was established at Clausthal, the smelting center, for education of officers of the Hanoverian mines at Schemnitz in Hungary. In the war of 1866 the district passed into the hands of the Prussians who improved the region's mines and furnaces.

Rivaling the Upper Harz, was Saxony in northern Germany between the Baltic and North seas on the north and Bohemia and Bavaria on the south. In this province were four metalliferous districts: Altenberg with 31 mines; Marionberg with 48 mines; Schwarsenberg with 146 mines; and Freiberg with 98 mines. Discovered in the 12th Century, the mines of Freiberg ranked among the oldest and richest in Europe. Their history is the history of the art of mining. In their numerous shafts and galleries can be traced the progress made from the days when a pit, a pickaxe, and a bucket constituted a mine, to the steam and hydraulic period which lightened the labor of men and allowed them to follow a vein with relentless perseverance so long as it had length, breadth or thickness.

Like Clausthal in the Upper Harz, Freiberg became a center for ore reduction, and by the mid-1860s its smelting works was processing about 33,614 tons annually, producing $2,000,000 worth of silver, lead, copper, zinc, nickel, bismuth and arsenic. The smelting works also received ore from South America and Mexico.[29] An outgrowth of this industry was creation of a great mining school.

The Freiberg school began about 1700 with the proprietors of the mines in the neighborhood of Freiberg sending their sons or assistants to a certain wise surveyor in the city, who gave them instruction in the best manner of seeking and working veins. For several decades it was nothing more than a provincial school. Then in 1765 a royal order transformed the school into the "Bergakademie" of Freiberg and it became the largest of Germany's mining schools, mainly through the exertions of a former student named Abraham Gottlob Werner.

Abraham Werner left the academy as a pupil in 1769 and returned to it as a teacher in 1775. The fame of his lectures and field trips attracted students from all parts of the world — Swedes, Russians, Englishmen, Frenchmen, Spaniards, Italians and Poles — all flocked to hear his novel theories upon the new sciences of mineralogy and geology. In a few years a small school of mines, before unheard of in Europe, developed to the rank of a great university, where Prussia annually sent deputations to be educated there. Soon, distinguished men of science throughout Europe were studying the German language so they could enter the academy to hear the great oracle of geology. From the time of Werner until the 1860s, the prosperity of the academy was uninterrupted, with able men filling professorships. Some of the most famous scientific scholars of the day studied and taught there.

In 1861 the school was dependent upon the government, being labeled the "Royal Saxon Bergakademie." Professors were appointed and received their salaries from the crown; the rules and regulations dictated by the Minister of Finance. General supervision of the school was given to the Royal Board of Mines. The president of this board and the head of the academy was the "Oberberghauptmann" or "Chief Captain of Mines," who next to the king was the highest officer of the state in the mining department. Without studying at Freiberg and passing an examination at the end of the course, no man could be appointed to any responsible position in the mines and furnaces of Saxony.

Although the school was designed to educate officers and directors of the Saxon mines, any native of the kingdom who could pass the entrance examination and furnish certificates of character could be admitted to the academy. Upon payment of 50 thalers per year, the student could attend as many lectures as he chose. Special advantages, however, were offered those expecting to enter the Saxon service at expiration of their course. Not only were annual fees waivered for certain cases, there was governmental assistance in the form of stipends. A liberal sum was often granted to defray traveling expenses when visiting distant mines and furnaces.

Like everything German, the cirriculum was regimented. Students had to take a practical six month course in the mines before they were allowed to listen to the lectures. During that time they donned the miner's dress and went down into the mines with workmen at six o'clock in the morning and came back to surface again at midday. By actual experience they learned to drill, dig, and muck ore. In so doing, students acquired knowledge of the appearance and parts of a mine, and the nature and habits of veins. As hard as this work was, it prepared students for the lectures which they would soon hear.

At first the student was confined to one mine and required to go through all the preliminary work there. Afterwards he was sent to such other shafts as would increase his knowledge. If he took the regular course in the academy, he had four years of lectures and recitations upon lectures to go through. At the end of this time he he could pass his examination and begin a year of practical work in the mines and furnaces before being regularly admitted to government service. Considering the wretchedly small salaries which Saxon mining officers received, it seems strange that so many struggled to obtain them.

The privileges of the academy were open to any foreign student upon presentation of the usual credentials and payment of fees. The latter were: first, an entrance fee of 6 thalers ($4.50); second, an annual academic fee of 15 thalers; and third, honorary fees for the lectures. As there were from 10 to 30 lectures (of two per week each) lecture fees varied greatly. The fees also gave the student the privilege of working in the chemical or metallurgical laboratory.

Besides general courses in geology, mineralogy, chemistry, surveying above and below ground, assaying, mining law, drawing and drafting, students at the Freiberg academy had the choice of three fields of study: mining, metallurgy, or engineering. For those desiring mining, there were two courses on the art of mining, and two courses on general mechanics, and two more on mining machinery. For students of metallurgy there were chemical and metallurgiucal laboratories, and lectures on general metallurgy, and the metallurgy of iron. Instruction in blowpipe analysis was given, as were practical exercices in dry and wet assaying. They also had the added opportunity of taking a practical course at the two great furnaces near Freiberg, where all ore from district mines were processed. They could also visit mines whenever they wished; guides being provided free of charge. What students observed on a large scale in the mines, could be studied in the form of models in the academy's collection. A library rich in works on mining subjects and mineral collections of the first rank were at their disposal.

In the year 1859-60 there were 143 students in the academy. Of these 56 were Saxons and 87 were foreigners. Of the latter, 15 came from English-speaking countries (nine from the United States). Of the rest there were Prussians, Austrians, Poles, Hungarians, Russians, Swedes, Norwegians, Frenchmen, Swiss, Italians, Spaniards, Chilians, and Bolivians.[30]

Such were the educational background of the metallurgists who would revolutionize smelting in the Great Basin. The first of these men, Joseph Mosheimer and Charles Stetefeldt, experienced difficulties. Others, however, left lasting marks. The first to appear was Guido Kustel, a renowned chemist, mineralogist, and metallurgist. Born about 1817 in Galecia, Austria, Guido was educated at Freiberg, Saxony. He came to California in 1851, perhaps with Joseph Mosheimer, for they partnered shortly thereafter in a San Francisco assay business.

One of few trained metallurgists on the Pacific Coast, Kustel's expertise was quickly recognized and he was soon putting into action what he had learned at Freiberg. In summer 1858 he initiated the Tyrolean tub (Freiberg) amalgamation process at the Heintzelman Mine near Arivaca in what is today southern Arizona. He also built a battery of Catalonian furnaces and vasos to smelt silver-lead ore from that mine, as well as from the Patagonia Mine, which later became the famous Mowry Mine. Being interrupted by the Civil War, that enterprise never amounted to much. Guido had better luck in Nevada, however.

When the Washoe excitement broke Kustel went to Nevada, and was among the first to work silver ores. He started an amalgamation works on the Comstock using the tub or barrel amalgamation principle, pan amalgamation coming afterwards. He introduced the roasting and barrel-amalgamation process at the Ophir works, and was for a long time connected with metallurgical progress of that region.

In 1863 Kustel published his work on processes of *Gold and Silver Extraction*, which created a sensation. His next work was on *Concentration and Chlorination*, published in 1868. He followed that treatise with *Roasting of Gold and Silver Ores* in 1870. With Ottokar Hoffman, his nephew, he experimented with the lixiviation of ores in 1867-68. An inventor, Kustel acquired several patents: one on an amalgamator, another on a concentrator. And his method of blowpipe assay for quicksilver was recognized as the simplest and best, explained in all European works. As a metallurgist, Kustel was known in every mining camp on the Pacific Coast, called upon to start up works where people were having difficulty processing their ores. He would test the matter, apply the remedy, and leave, never taking advantage of the situation for personal gain.

In spring 1870 Kustel studied Eureka's smelting problems, and outlined five ways to improve productivity in the district. He suggested that ores be heap roasted to volatilize part of the arsenic and sulphur, a process that should be done at the mines where wood was cheaper than in Eureka. More careful burning of charcoal was advocated to obtain hard, larger pieces. Since this fuel was delicate, subject to crushing during freighting to the smelters, its transportation should be done in racks as in Pennsylvania and New York. Kustel also advocated a change in smelting techniques.

To change the composition of slag, regular charging was suggested. Before going into the furnaces, ore must be carefully mixed on the charge-floor with silver-bearing quartz from New York and Secret canyons. No slag should be mixed with carbonates. This would result in a slag lying between a singulo and bi-silicate. Such a slag was hot, light, and not inclined to stiffening, thus would prevent mechanical loss of lead, assuring longer life for furnace walls and crucibles. Most important, Kustel asserted smelters "must be constructed differently."

"Walls must come down straight to the hearth, or contract gradually about one foot in the whole height from top to bottom, like the Raschette or the Piltz furnaces; the mouths of the tuyeres ought to be narrowed from three to one and a half inches, and pressure blowers ought to be employed instead of the fan-blowers. . . . The fore-crucible ought to be closed, so as to protrude not more than four inches from the breast at the commencement of the campaign, and not wider than six inches. As long as no dust chambers can be connected with the furnaces (which would undoubtedly be the best), the stack ought to be rather wider than the furnaces than narrower, so that the draft may be reduced to a minimum, and thus the escape of dust be prevented as much as possible."[31]

Guido's suggestions were acted upon by three metallurgists: Albert Arents, Otto H. Hahn, and Winfield S. Keyes. We know something about the latter gentleman, he being the hard-nosed superintendent of the White Pine Smelting Company. In 1871 Keyes assumed the superintendency of the Eureka Consolidated Company's smelters. The other two, strangers to Eureka, should be introduced.

Born in 1845 at Appolta, in Thuringia, Otto H. Hahn graduated from the Bergakademie at Clausthal in the Harz. He was also a graduate of the University of Jena. Upon completion of his studies, he came to America in 1963 and found employment in New York City with Adelberg & Raymond, doing some work in connection with New York lead mines. Hoping to find a broader field for his activities, Hahn ventured to California where he fell in with Charles A. Stetefeldt. After working with him for a while, Hahn went to Austin, Nevada, to take charge of Stetefeldt's assay office. Later he moved to Virginia City, Nevada, where he engaged in assaying and mining for several years, during which time he was associated with Luckhardt, another metallurgist.

Albert Arents was born in Germany about 1840. Receiving his education as a mining engineer and metallurgist at the Royal Schools of Mines at Clausthal and Berlin, he served his "practical course in metallurgy" from 1859 to 1864 in Harz reduction works. Coming to America, Arents found employment in early 1866 at the copper smelting works of Crane & Brigham at Hunter's Point near San Francisco. Four years later he came to Eureka as metallurgist for the Eureka Consolidated Company.[32]

In spring 1870 superintendent Winfield S. Keyes and metallurgist Albert Arents commenced a program to increase the profitability of the Eureka Consolidated Company. Following Kustel's advice, Keyes and Arents overhauled the company's reduction works, replacing the small square shaped shaft smelters with furnaces of higher capacity. They chose the German designed Piltz or Pilz furnace, which was a modification of the round furnace introduced by the English into Spain about 1843.

First constructed in 1867 at Balsbrucke, near Freiberg, Saxony, the Piltz furnace was octagonal shaped in its horizontal section; the octagon being flattened at the front or working side of the furnace. Supplied with seven narrow mouthed tuyeres, the Piltz furnace was capable of producing temperatures of 2,000 degrees. While the Piltz furnace processed more ore, up to 24 tons in 24 hours, and produced more bullion, it was not trouble free. Nevada ores, with their high percentage of gangue, still formed bridges and sows in the furnaces, requiring use of bars to keep the slag flowing. To correct that problem the Piltz furnace design had to be altered.

Arents remodeled the furnaces by elongating the smelter from back to front, giving it parallel sides, and rounded corners at the

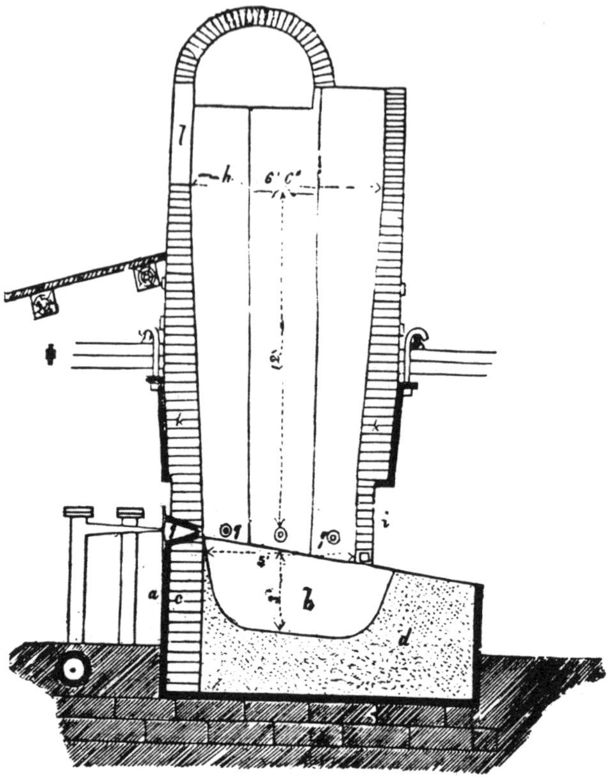

PLATE II, FIGURE 1.—SECTION.

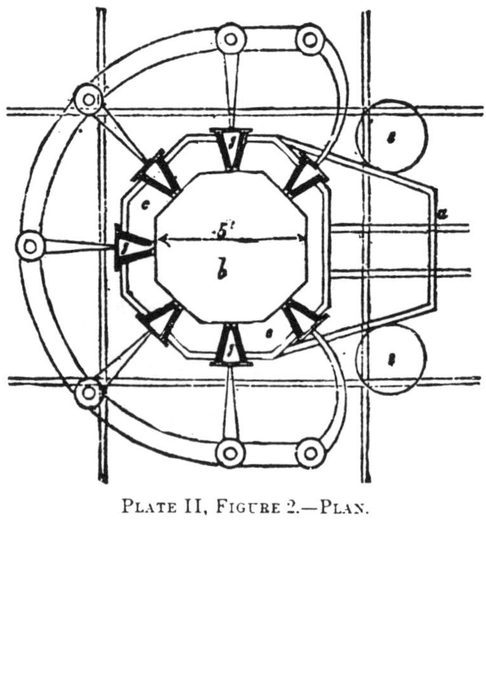

PLATE II, FIGURE 2.—PLAN.

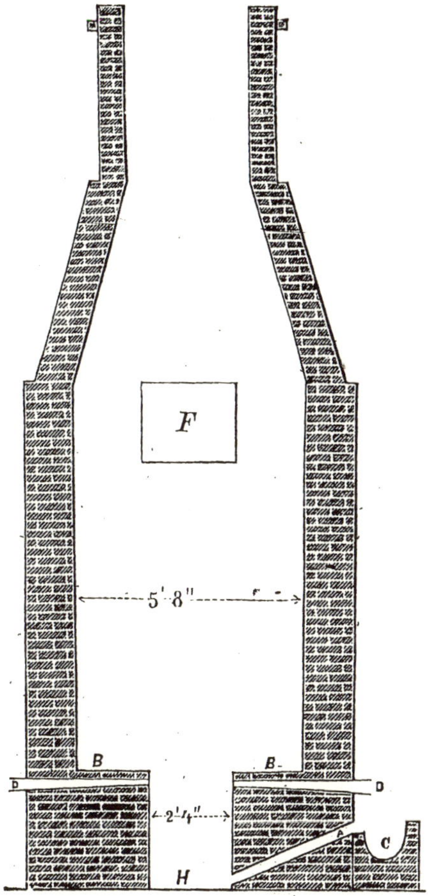

Above: diagram of Bassett's smelting furnace, which operated at the Silver West Smelting Works, Eureka, Nevada. Patterned on the Raschette plan, this furnace had 14 tuyeres. (*Mining and Scientific Press,* August 10, 1872)

Left: The Piltz furnace originated at Freiberg, Saxony. Its horizontal section was octagonal; the octagon being flattened at the front or working side of the furnace. It was supplied with seven tuyeres. Plat II, Figure 1, shows a section from front to back, and Plate II, Figure 2, shows the ground plan. (*Annual Report,* California State Mining Bureau, 1888)

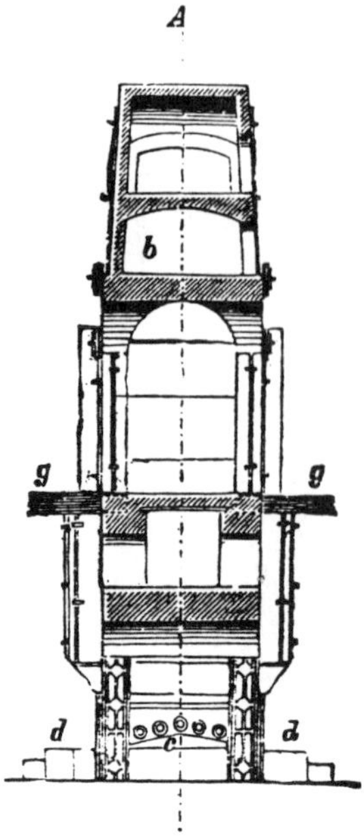

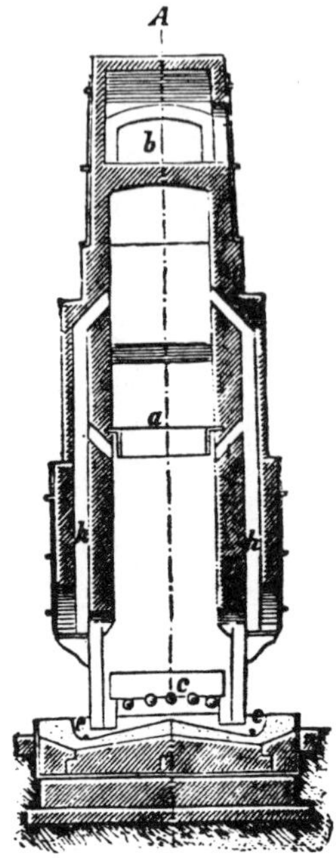

PLATE III, FIGURE 1. PLATE III, FIGURE 2.

The Raschette furnace as installed by Arents at the Eureka Consolidated. As originally constructed this furnace had two working fronts, and was proportionately much longer than it was wide. Figure 2, Plate III, represents a section through the working doors. Forty percent more ore was put through this furnace in 24 hours than round furnaces. The Raschette furnace formed the basis of construction of all modern American smelters. (*Annual Report*, California State Mining Bureau, 1888)

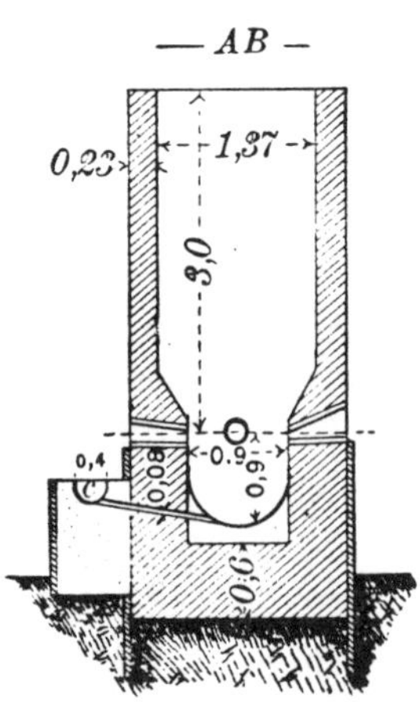

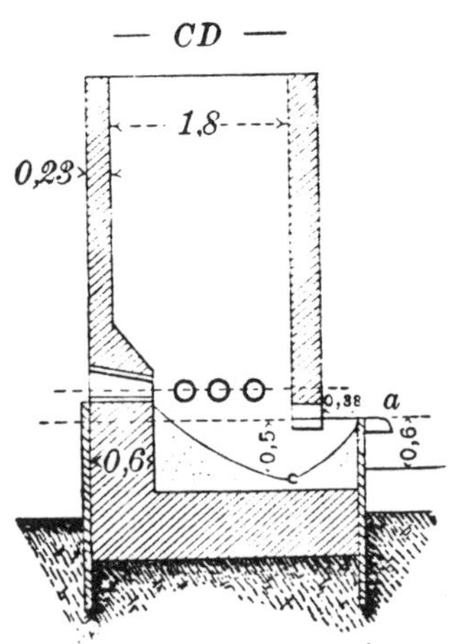

PLATE IV, FIGURE 1. PLATE IV, FIGURE 2.

back. He also narrowed the hearth. With exception of having only one breast, the new furnaces came very near the shape of the Raschette or Rachette furnace.

Sometimes called an Alger furnace, the Raschette was invented by Vincenz von Raschette, Major-General of the Russian Mining Engineers, to smelt copper and iron ore at Prince Demidoff's works in Russia. In 1864 Charles Aubel introduced the Raschette furnace at the Altenau Smelting Works near Clausthal, in the Upper Harz region of Germany. Used to smelt both copper and lead ores, this furnace differed from other blast furnaces, its horizontal section being long and rectangular, and having a row of several tuyeres on each of its long sides, with a tap hole at each end or narrow side. The result was a furnace pictured in the accompanying sketches.[33]

While working on the furnaces of the Eureka Consolidated Company, Keyes and Arents came up with "one of the most important inventions ever made in the construction of smelting furnaces," the siphon tap, or lead-well.[34] Also called the "sub-surface inverted siphon tap," this device was no more than a pipe running at an angle through the lower portion of the furnace wall at the bottom of the crucible, from which it slants upwards ending in a kettle, the upper rim of which was at a level with the height of the lead-bath in the furnace, when the crucible was full. As the metal rose in the furnace, it also rose in the tube, overflowing into a kettle at the discharge end of the pipe. The result was a tapping method which produced cleaner bullion.

This arrangement of tapping was continuous, carrying the molten lead out of the bottom of the furnace as fast as the metal accumulated. At the same time the charge above remained in the crucible long enough to give the molten ingredients the required time to react with each other, and separate according to specific gravity. Because the siphon tap never drained the crucible, lumps of iron and other impurities, the constituents of sows, always floated on the surface of the lead bath where they were exposed to the oxidizing influence of the blast. What remained of the iron was eventually carried away in the slag. Thus formation of "sows" was prevented and furnace life prolonged. Lightening the work of smelters, the siphon tap was adopted by all Eureka furnaces, as well as lead refineries throughout the West. As patentees of the siphon tap, Arents and Keyes collected $250 a year from every furnace utilizing their invention.[35]

For four years every smelting company at Eureka either altered their furnaces or built new ones. In summer of 1870 Otto H. Hahn and his assistant, Karl von Leibenau, planned and constructed the smelting plant of the Richmond Company. Their first furnace was six-sided, with five tuyeres, and was capable of smelting 24 tons a day. Other furnaces of the same design followed. In summer of 1872 the Silver West Smelting Works erected a new form of smelter designed by a Mr. Bassett. The novelty of its construction consisted in the arrangement of tuyeres, the form of the lower portion of the furnace, and the manner of drawing off the lead. Without a doubt the years 1870 to 1874 was a period of great innovation at Eureka, much to the benefit of American smelting.

According to the *Mining and Scientific Press,* by 1877 Eureka had attained the "highest degree of perfection in the world of smelting." With exception of two water-jackets at the Richmond Company, all furnaces at Eureka were modifications of the Raschette and Piltz furnaces, altered and improved to suit the combinations and qualities of the ore smelted.

This article, reproduced from the *Eureka Sentinel,* presents the best description of the type of furnaces then being erected at Eureka.

> These furnaces were built of stone from quarries in the vicinity, and lined with a porous fire-proof sandstone, of excellent quality, from Pancake mountain, twenty miles distant. The inside diameter will average $4^1/_2$ feet at the hearth, and the height to the charging windows is about 16 feet. The hearth rises about $2^1/_2$ feet from the ground and projects 18 inches from the mouth in the front of the furnace. On three sides and one foot above the hearth, eight tuyeres pierce the shaft, three on each side and two at the back. The shaft is carried through the floor and the charging windows open on the second story, or ore floor of the works, but the masonry extends to and through the roof, connecting with the flue or fume arrester. The tuyeres are double wrought iron pipes, a space intervening between the outer and inner pipe, through which a constant current of air flows, preventing the melting of the tuyeres

Right: This smelting plant, typical of those erected at Eureka, consisted of a shaft furnace with metallic bosh or water-jacet in section from front to rear, showing five tuyeres on one side; also the blower and the downcast pipe connecting the upper part of stack with dust chambers in the rear of the blower. The upper portion of the furnace rested on stout iron columns. The skeleton building surrounding the same is also shown. The Baker blower is represented immediately to the rear of the furnace. (*Annual Report,* California State Mining Bureau, 1888)

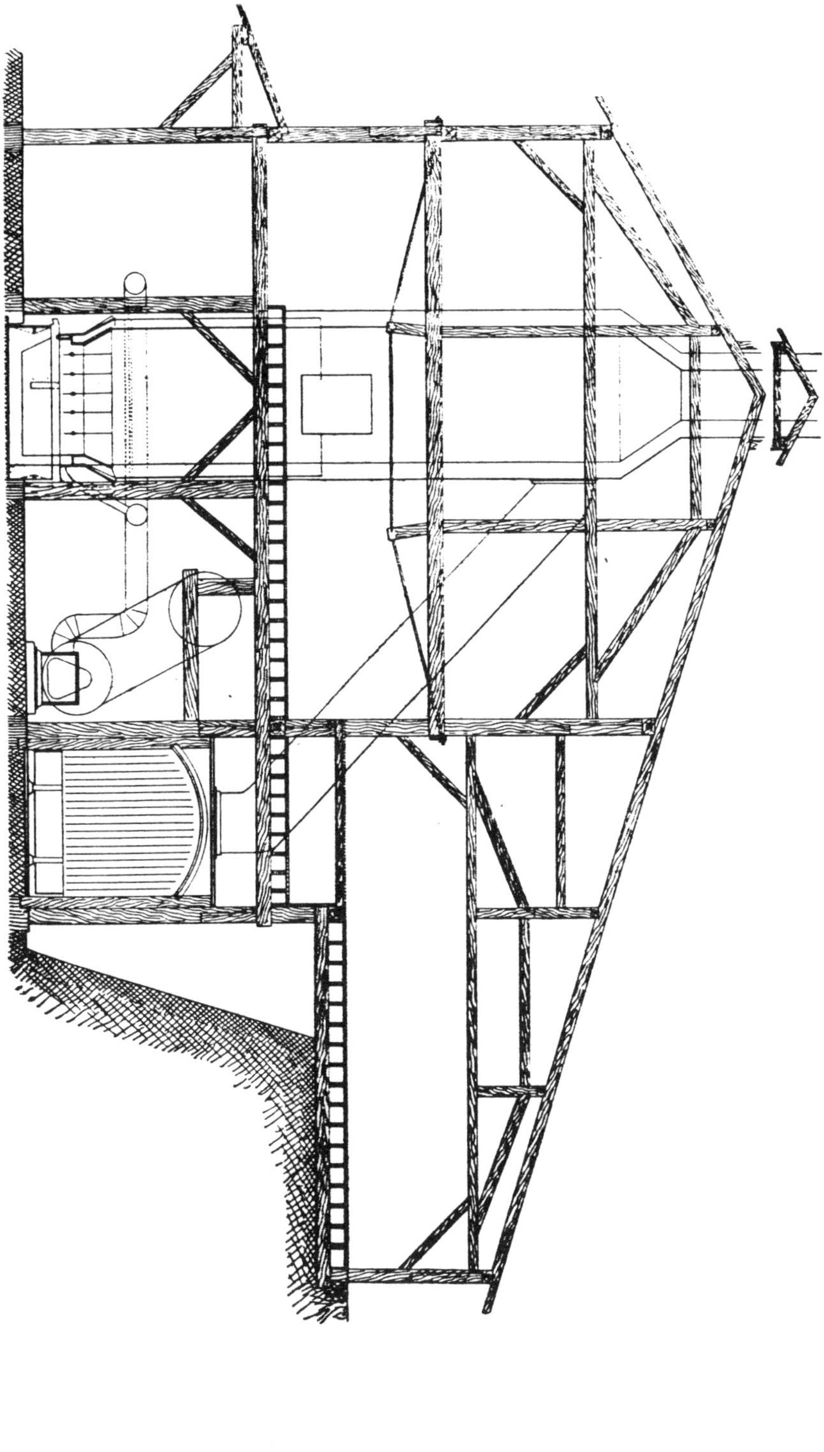

where they are exposed to the fierce heat. The blast is furnished by a Roots blower, with a capacity of 600 revolutions per minute, and is conveyed to the tuyeres by a system of tin pipes and canvas hose. The front of the furnace is open to a height of 10 inches, the space between the melted slag and masonry being temporarilly filled with a mixture of fireproof clay, and removed at the pleasure of the smelter, when it becomes necessary to "bar out" or remove the slag. At the side is the lead well, connected by a pipe with the bottom of the furnace, the lead rising to the same height in the well that it attains in the furnaces. The projecting hearth is supplied with two spouts, through which the iron and slag are drawn off into the cast-iron slag-pots. In the working process the furnaces is filled with the ore mixed with a certain proportion of charcoal. Fire is applied, the blast turned on, and the furnace is in operation.

The ore smelted is argentiferous galena, and contains from 40 to 60 percent lead, some iron, sulphur and arsenic. The heat concentrated on the ore at the mouth of the tuyeres, will reach 2,000 degrees Fahrenheit, and under its influence the metals contained in the ore and the silica or quartz melt and flow to the bottom of the furnace. The lead, from its greater gravity, goes to the bottom, carrying with it the precious gold and silver. On the top of the lead the melted iron floats, and on this again swim the lighter materials termed slag. As the lead accumulates it is ladled out from the well into molds, forming base bullion bars, averaging 120 pounds in weight. These bars will assay from $200 to $400 per ton, varying with the quality of the ore smelted. Each charge is composed of 46 shovelfuls of ore, intermixed with 12 bushels of charcoal. No flux is necessary, with the exception of a shovelful of silica or slag, as the ores smelts freely. One of the furnaces at the Eureka Consolidated, now running, was built for a reduction capacity of 50 tons per day, but it goes considerably above this, averaging 65 tons every 24 hours, and has reduced as high as 90 tons, producing 10 tons of bullion.

The working force consists of one smelter, two slag-wheelers, two feeders and three ore and charcoal tenders. The smelters receive $5 per day of eight hours, and wheelers $4. The work is rather unhealthy—the most exposed to the fumes being subject to lead paralysis; but of late the attention paid to ventilation has remedied this in a great degree.[36]

An account of how the smelters were manned and operated was published by *Eureka Sentinel*. Picked up by the *Mining and Scientific Press* of April 14, 1877, it compliments the above description.

The working hours at smelting furnaces . . . are divided into the same periods as at the mines — eight hours — and they change their hours every two weeks. Each furnace employees a force of 15 men (exclusive of ore and charcoal men), of whom there are three smelters, six feeders and six slag-wheelers. The wheelers are employed on the second or feeding floor of the works; their labor consists of charging the furnaces with the requisite amount of coal and ore through the feed holes of the shaft. It requires no special training to do the work, but the men are exposed to the ascending fumes, and it is generally looked upon as the most unhealthy part of the work. The slag-wheelers are kept busy hauling away the melted debris, iron and bullion, and assisting the smelter whenever occasion may require. The work is very laborious, and eight hours is sufficient to exhaust the physical powers of the strongest man.

The smelters' duty lies entirely in front of the furnace, and it is to his care and attention that its success is due. Green men will not answer for this part of the work and their ranks are recruited from the slag wheelers, whose ambition leads them to aspire to the position. Their work consists of watching and assisting the flow of slag that floats on the surface of the metals, which consists of liquid silicia, limestone and the impurities of the ore, tapping and drawing off the iron when it accumulates to a certain height in the crucible, ladling out the base bullion from the lead well into molds, regulating the blast and testing the iron and slag to see that no lead is escaping from the spouts. One hour after commencing their work the blast is shut off and the whole force congregate in front of the furnace preparatory to "barring out." This, however, is only the rule in the shaft furnace, the hydrocicles [water-jacket furnaces] dispensing with the operation. It is a vexed question as to whether anything is gained in the water-jacket by this improvement, but we will not discuss the arguments pro or con. There is no doubt that in the shaft furnace, the operation more than compensates for the time lost in performing it. We have described the work at some length before, and will not now attempt a description of it in detail. It consists in removing the chilled slag and iron that has cooled outside the hearth, breaking down the old and putting in a new clay front, and stirring up the inside so that it is loose and free from horses, thus facilitating the smelting of the ore. It generally consumes from a half to three-quarters of an hour, and on its completion the blast is turned on and the work proceeds without interruption during the remainder of the shift. The work varies in laboriousness, and depends greatly on the quality of the ore.

> Ores with a large percentage of lead, as a matter of course, form greater quantities of bullion, while other ores will run to slag and iron. The greater the quantity of slag the more work for the wheelers and feeders, and vice versa. The wages received are fair, the smelters getting five and the others four dollars per day for their services. Notwithstanding the prevailing impression that the work is very unhealthy, there are always a surplus of applicants for vacancies, and no trouble is experienced in procuring help. Some men are more susceptible to the influence of the fumes, and soon succumb to its poisonous effects, while others work around the furnaces . . . without experiencing any deleterious symptoms. It is noticeable that the men who pay thorough attention to cleanliness and dissipate the dust are those who escape. There have been great improvements made in the way of ventilation and this has caused a marked decrease in unhealthiness of the work. As a class those employed are intelligent and industrious and there are many who have followed the occupation ever since smelting was inaugurated at Eureka.[37]

There may have been a surplus of applicants for work at Eureka's smelting works, after all most men concided that "fumes would not bother them if they were hungry." The fact remained that none but the strongest could stand the labor. The *Eureka Sentinel* related the situation between shifts at the smelters:

> Only a few nights ago, when the 11 o'clock shift was whistled, there was no less than 25 men standing about the Richmond furnaces and refinery. If a man gets leaded and fails to show up when time is called, some man from among those seeking jobs or standing about, is at once put on in his place, and the "falling down" is of such frequent occurrence that the men congregate there to get a job if possible. There they are by the score at every change of shift.
>
> On the occasion to which we referred, a rather small man was put on. He did not look very strong. His job was to wheel out a slag-pot from the refinery. George Stitt, a powerful man, who has walked from Secret canyon over Prospect mountain, and as far northeast as the Newark mines, in search of work, and who has not missed to be on the ground at a change of shifts at some of the furnaces for upwards of six weeks, was standing by when the little one went to work.
>
> George communed with himself, and the conclusion arrived at was that he (the young fellow) could not stand it to work the shift through. All other seekers of jobs had departed.

> George took up favorable position to see the young man tumble and be on hand with his services if the foreman should so order.
>
> In George's words: "The first hour the little chap worked like a bugger; about 12:30 his returns with the trucks were markedly slow; the water was running from every pore in his hide, and his hand would sweep over his face to be filled with perspiration. At 1 a.m. I says I'll have him in 30 minutes — he was as pale as a ghost and was trying to vomit. I stayed till 3 o'clock and the little one was still moving along the track, spilling slag all over the same, but he had the d—dest grit ever stuck in so little hide."
>
> "The 'little one' worked a couple of shifts, but went by the board through pure lack of strength to perform the work allotted him and the fumes he had to take."[38]

The atmosphere about the furnaces, contaminated with lead and arsenic, may have killed off men, but smelter construction concentrated population. W. W. McCoy and Alonzo Monroe surveyed a townsite adjacent to their smelter in 1869, and two other townsites were made by O'Neil and McDonald. In July 1870 the district contained a thousand inhabitants, 600 to 700 of whom had settled where the smelters were located, in a mile and a half long north-south trending canyon between projecting spurs of the Prospect Mountains.

While there was water and fuel, lumber for building purposes was in such short supply that it had to be imported from Truckee. Consequently many of the town's first buildings were hauled in from Hamilton, Shermantown, Elko, and Carlin. By August the town boasted of some 400 houses and the population had increased in 45 days to 2,000. Under a disagreeable canopy of smoke and fume the town of Eureka sprung to life.[39]

Eureka's first link to the outside world was the freighting line of Seaton & Clark, whose ten-mule teams brought in provisions, equipment, and supplies, and freighted bullion to Carlin, at a charge of $30 per ton. In early 1870 John A. Wilson's semi-weekly stage line between Austin and Hamilton was changed to pass through Eureka. This line brought in mail, which Eurekans picked up at the town's first post office, established in the Parker House opposite the Eureka Consolidated Works. By summer of 1870 there were two other stage lines: Woodruff and Enner running to Hamilton and Palisade station, and Beachey, Wines & Company connecting with

Elko. The latter stage line brought in an average of 90 passengers a week. A host of restaurants, stores, boardinghouses, blacksmith shops and wagon works catered to needs of the community and visitors.[40]

In 1874 the Eureka and Palisade Railroad Company commenced building its narrow gauge line from Palisade to Eureka, a distance of 90 miles. The line was completed in October 1875 and the "Slim Princess" puffed into town. Advent of the railroad made Eureka the depot for all wagon transportation, freighting and passenger traffic for numerous mining camps to the south. Eureka was now in regular stage communication with Austin, Belmont, Tybo, Hamilton, Pioche and the Ward District.[41]

Next to mining, Eureka's most important industry was production of charcoal. The district's furnaces consumed 30 to 35 bushels to every ton of ore smelted. In 1871 the five furnaces of the Eureka Consolidated Company consumed 4,600 bushels daily. That was a lot of wood, considering that a cord of nut pine produced about 28 bushels of charcoal. As more furnaces were added the annual consumption of fuel soared, exceeding 1,200,000 bushels by 1876. The charcoal trade had become big business, furnishing work for 2,000 men, prinicpally Italians, who had drifted in from California and the southern states, bringing with them European techniques of burning charcoal in heaps. They worked in teams; some cutting trees and sawing branches into proper lengths, others loading and teaming the cord wood to burning areas close to the smelters.

Prior to 1872, when fuel could be harvested close to Eureka, the price of charcoal hovered around 15 cents per bushel and remained fairly constant for four years. But as the harvest went on, the perimeter of nut pine, mountain mahogany and cedar retreated. In 1876 the average distance between the smelters and point of harvest had reached 35 miles.[42] With diminishing of fuel resources the price of charcoal crept above 20 cents a bushel, and Eureka smelters began eyeing the forests of the Sierra Nevada and the Wasatch Mountains. More on that in a later chapter.

Innovation brought consolidation. Within four or five years the number of smelting plants at Eureka went from 12, each charging a different price to process ore, to two large concerns: the Eureka Consolidated and the Richmond Consolidated. By 1879 the latter company's furnaces, with the exception of the water-jacket ones, each had a capacity of reducing 75 tons of ore a day. Combined, the companies smelted 200 tons of ore daily. About the furnaces 150 men were employed; 200 more worked in the mine. The Eureka Consolidated ran four furnaces in 1879; one furnace, plagued by trouble, being shut down.[43]

Eventually the Richmond and Eureka Consolidated combined to fix a smelting charge which was "all the miners could stand and just low enough to prevent the ore going to Salt Lake City." In 1884 the price of lead declined, and Eureka suffered like every lead-producing district in the United States. The Richmond and the Eureka Consolidated companies shut down their mines and idled furnaces. By May 1885 the *Mining and Scientific Press* reported that "at no former period in the history of Eureka have times been as quiet and dull as at present."[44]

The Eureka Consolidated Company used the opportunity to modernize its reduction facilities. In early 1885 it tore down the old furnaces and erected a new water-jacket smelter. Fabricated by the Pacific Iron Works of San Francisco, this 60-ton capacity furnace, which was blown in on May 22, 1885, was a combination of the Raschette and Piltz furnace. A second smelter, "a facsimile" of the first, was put up that summer. Designed by Frank Robbins, son of G. Collier Robbins who constructed the first smelter at Eureka in 1869, this reduction works was designed for utmost efficiency, with self-dumping cars drawn over a cable tramway to the furnace feed floor. Everything about the works was so arranged that no shoveling of ore was required from the time it was mixed until it was fed to the furnace.[45]

Not to be outdone, the Richmond Company fired up two of their furnaces in late spring of 1885, which caused the *Eureka Sentinel* to comment that "the business air of the community again assumes tangible form."[46] The recovery was shortlived for a number of reasons. The price of silver had been dropping since demonetization in 1873, while price of fuel and transportation had climbed, and exorbitant charges and discounts by the two large reduction works stiffled independent mine production in the district.[47]

If that was not enough, the Eureka district suffered a "bad setback" in 1888 when the government filed suit against the Richmond Consolidated Mining Company for illegally removing timber from public lands.[48] Initiated by an "impetuous" attorney general, the suit centered on interpretation of the Timber Act of 1878 which permitted bona fide residents to cut trees on mineral land for purposes of mining, agriculture, or other industries. The action claimed that between 1873 and 1888 the Richmond Mining Company, a British corporation, "had cut and removed pine, cedar, and mountain mahogany totaling no less than 21,600,000 bushels of charcoal and 112,320 cords of fuel wood," with an estimated value of $6,246,720. The case was one, contended the Commissioner of the General Land Office, in which a "foreign corporation had appropriated to themselves the liberal provisions of the Government made for the benefit of its citizens only."

In response to the Federal action the Richmond Consolidated Mining Company at end of 1889 closed its mines, throwing many tributers out of work, and shut down its furnaces. Angry protests followed from British stockholders. "In the face of these suits," warned the directors, "it behoves the Company to be very careful how they invest any further capital in the United States." This caused the Nevada legislature to petition Congress, asserting that business was paralyzed "and the people are menaced with absolute financial ruin." The Senate Committee on Mines and Mining investigated and found the timber suit "needless and vexatious." In authorizing them, the attorney general "must have done so under a miapprehension of the facts." As a result, after a test case was decided in November 1889, the charges were dismissed, much to the relief of British shareholders.[49]

In January 1890 the Richmond furnaces were running again, and Eureka Consolidated started up its smelter the first week of June 1891. The glory days did not return, however. With the playing out of the great lodes of argentiferous lead, miners were laid off at the Richmond, Eureka Consolidated, and other mines. At end of September 1891 the Eureka Consolidated made a "cleanup" of ore at their mine, the last to be worked at their furnace. From then on the company shipped low grade lead ore to Salt Lake City, the result was that the district's population "thinned out," and Eureka shriveled and almost died.[50]

* * *

1. Hubert Howe Bancroft, *History of Nevada*. San Francisco, 1890, pp. 277-28.

2. Lead was worked for $160 per ton in New York, and $150 in San Francisco; and with increased facilities of the continental railroad, returns were made to the smelter from separating works in either section of the country in a much shorter time than from Swansea. *Mining and Scientific Press*, April 23, 1870.

3. *Ibid*., April 10, 1869; June 11, 1870.

4. *Ibid*., April 10, May 15, November 6, 1869.

5. J. Mosheimer, "Ore Smelting at White Pine District, Nevada," *Ibid.*, June 11, 1870, p. 401.

6. *Mining and Scientific Press*, April 16, 1870.

7. Francis M. Roots, who was largely responsible for the invention of the Roots blower, was born in Oxford, Ohio, on October 28, 1824, a son of a well-to-do farmer. Receiving his education in Miami University, he moved in 1845 to Connersville, Indiana, to establish a large woolen mill, which he and his brother conducted until 1859. That year Francis disposed of his interest in the business and entered upon a career as an inventor. A year later he perfected the world's first successful positive blast rotary blower. Roots also invented and manufactured the Roots' gas exhauster, known to the gas industry as the Roots' steam pump. He did not solely confine himself to the manufacture of his inventions. He was president of the First National Bank of Connersville, of the Connersville Furniture Company, and in several other businesses in Connersville and Cincinnati. Francis M. Roots died on October 25, 1889, at his home in Connersville, Indiana. For additional details on the life of this brilliant man see *Scientific American*, February 28, 1880, p. 130; April 3, 1886, p. 217; January 4, 1890, p. 6; and the *Engineering and Mining Journal*, November 2, 1889, p. 390.

8. *Mining and Scientific Press*, November 6, 1869.

9. *Ibid.*, April 23, 1870; J. L. Jernegan, "Lead and Silver Smelting in Chicago," *Transactions*, American Institute of Mining Engineers, Vol. II, 1874, pp. 279-95.

10. After a long and prosperous business career, Edward Balbach died honored and respected in Newark, October 14, 1890, leaving his son, Edward, Jr., in charge of the extensive establishment. Edward Balbach, Jr., died at Newark on December 30, 1910. *Engineering and Mining Journal*, March 28, 1891, p. 375-76; January 14, 1910, p. 116.

11. *Ibid.*, January 12, 1907, p. 106.

12. *Mining and Scientific Press*, September 24, 1870.

13. *Ibid.*, October 1, 1870; January 14, 1871.

14. According to the *Mining and Scientific Press* of September 24, 1870, "only three furnace proprietors out of more than twenty that have operated, have even paid for the ore which they have received from the miners. Some of the furnace men, after accumulating huge stocks of silver-lead bars, have shipped them away to Newark or San Francisco, and then followed them to obtain the returns, leaving the miners and operatives to whistle for their pay and keep their courage up in the same manner."

15. *Ibid.*, October 1, 1870.

16. Myron Angel, *History of Nevada*. Oakland: Thompson & West, 1881, p. 434.

17 *Mining and Scientific Press*, September 30, 1865; *Engineering and Mining Journal*, July 21, 1894, p. 59; March 24, 1930, p. 310; *Transactions*, American Institute of Mining Engineers, Vol. XXVI, 1896, pp. 537-544.

18. In 1870 Stetefeldt returned to Europe intending to settle permanently. But his father had died and the friends of his youth were scattered. Realizing that his American ties were stronger, Charles returned to San Francisco in 1872. He resided on the Pacific Coast until 1882, and then moved to New York, remaining there until 1889, at which time he moved back to California, taking up residence at Oakland. During much of this time he consulted with mining companies, experimented with ore reduction, and improved the Russell lixiviation process. He produced numerous technical articles, 21 of which were published in the *Transactions* of the American Institute of Mining Engineers. Although he did not join that organization until 1881, he became an active member, being elected a vice-president of the Institute in 1888, and again in 1895. In summer 1894 Stetefeldt gave up smelting, roasting, reduction, assaying, and other metallurgical pursuits for more attractive and less responsible research in speculative philosophy, ethical and physical sciences. In July of that year he read a paper before the Astronomical Society of the Pacific Coast in answer to the problem, "Can Organic Life Exist in the Planetary System Outside of Earth?" Charles Stetefeldt died of heart disease at Oakland, California, on March 17, 1896. Although he married in December 1872, Charles had no children; his wife died some years before him. *Mining and Scientific Press*, September 30, 1865; *Engineering and Mining Journal*, July 21, 1894, p. 59; October 26, 1912; March 24, 1930, p. 310; *Transactions*, American Institute of Mining Engineers, Vol. XXVI, 1896, pp. 537-544.

19. W. W. McCoy was educated as a physician, but when the Mexican War broke out he joined the volunteers and rose to the rank of major of an Indiana regiment. He fought from

Vera Cruz to the capture of the Mexican capitol. Returning to Indiana, he served a term in the state legislature, and then went to California, settling in Santa Clara County; he represented that county in the California legislature. He was also active in Nevada politics, being a state senator from Eureka in the 1875 session. He was also the democratic caucus nominee for United States senator against J. P. Jones. He died in Shasta County, California, about July 1, 1881; aged 70 years. *Salt Lake Tribune*, July 10, 1881.

20 Angel, *History of Nevada*, pp. 425-26, 429-431; *Engineering and Mining Journal*, October 26, 1912.

21. Angel, *History of Nevada*, p. 288.

22. Buel and Bateman sold one-half of their share in the smelting works for $12,000. *Mining and Scientific Press*, July 9, August 13, 20, 1870; *Engineering and Mining Journal*, September 20, 1870.

23. *Engineering and Mining Journal*, September 20, 1870.

24. These claims, some patented and other not, consisted of the original Richmond and Tip Top, the Lookout, Victoria, Silver Region, Colorado, St. George, St. Patrick, St. David, St. Andrews, Standard and Cyrus. In 1872 H. Heynemann, who purchased the Dunderberg and other mines, built two furnaces, and that year W. S. Keyes, superintendent of Eureka Consolidated, located a group of claims adjoining the Eureka Consolidated on the east, and formed the K. K. Consolidated Mines. Its ore was processed in a furnace built by the Silver West Mining Company in 1872. Angel, *History of Nevada*, p. 433.

25. Ore prepared for smelting on an average contained 40 to 48 percent lead, $60 to $80 silver, and $15 to $20 gold per ton. In bulk the ore, which also contained arsenic, was a yellow color, due to its iron content. When smelted the arsenic combined with the iron to form a white compound (speise) somewhat like matte, holding $36 to $45 per ton in silver and gold, and 24 percent lead. The ore smelted readily by itself, nothing being added except about 20 percent slag.

26. *Salt Lake Tribune*, July 4, 1871.

27. *Mining and Scientific Press*, August 27, 1870.

28. *Ibid.*, August 13, 1870.

29. William P. Blake, "Something About Freiberg Mines," *Mining and Scientific Press*, October 12, 1867, p. 226; Anonymous, "The Mines of Freiberg in the Olden Times," *Scientific American*, Vol. IV, No. 20, May 18, 1861, pp. 316-17.

30. Anonymous, "Mining Schools of Germany — The Academy at Freiberg, Its History, Present Condition and Peculiar Advantages," *Scientific American*, Vol. IV, No. 21, May 25, 1861, p. 326.

31. The suggestions of Guido Kustel can be found in the *Engineering and Mining Journal* of September 20, 1870. Kustel's activities at the Heintzelman Mine for the Sonora Exploring & Mining Company are detailed in Diane M. T. North's *Samuel Peter Heintzelman and the Sonora Exploring & Mining Company*. Tucson: University of Arizona Press, 1980. At the time of his death, on August 16, 1882, Kustel was maintaining an assay office in San Francisco with his only son as partner. His death was attributed to "inflammation of the kidneys," later known as Brights Disease. *Mining and Scientific Press*, August 19, 1882; *Engineering and Mining Journal*, September 16, 1882, p. 145.

32. *Engineering and Mining Journal*, October 10, 1885, p. 255; October 24, 1885, p. 287.

33. See *ibid.*, October 10, 1871; John Percy, *The Metallurgy of Lead, including Desilverization and Cupellation*. London: John Murray, 1870, pp. 386-387. For contribution of Vincenz von Raschette see *Engineering* (London), Vol. XVII, January 30, 1874, pp. 83-84.

34. S. F. Emmons, and G. F. Becke. *Statistics and Technology of the Precious Metals*. Washington: Government Printing Office, 1885, p. 660; *Mining and Scientific Press* of May 13, 1871; *Tribune*, July 4, 1871. While the siphon tap was universally used, Arents and Keyes had difficulties collecting royalties. They waged a five year fight to collect fees from the Grant Smelting Company of Denver which had 13 furnaces. In the end they were victorious. *Mining and Scientific Press*, October 9, 1886, p. 232.

35. *Mining and Scientific Press*, February 10, 1877, from an article published in the *Eureka Sentinel*.

36. *Mining and Scientific Press*, April 14, 1877.

37. Reprinted in *ibid.*, February 19, 1881.

38. *Ibid.*, July 9, August 13, 1870.

39. Angel, *History of Nevada*, p. 439.
40. *Ibid.*, p. 440.
41. Guido Kustel, "Smelting at Eureka, Nevada," *Mining and Scientific Press*, August 20, 1870; O. H. Hahn, "Notes on Lead Smelting in Nevada and Utah," *Tribune*, November 29, 1871.
42. Angel, *History of Nevada*, p. 431, 438.
43. *Tribune*, February 25, March 25, 1879.
44. *Mining and Scientific Press*, 2, 1885.
45. *Ibid.*, January 24, May 2, 1885, pp. 61-62; June 6, 1885, this article gives specifications of new smelter; *Eureka Sentinel*, Feb. 21, 1885.
46. *Sentinel*, May 23, 1885.
47. *Mining and Scientific Press*, September 27, 1890, p. 201.
48. *Ibid.*, February 2, 1889, p. 79; March 16, 1889, p. 182; March 30, 1889, p. 222, 232; April 6, 1889, p. 245.
49. Clark Spence, *British Investments and the American Mining Frontier*, pp. 196-197; *U.S. Senate Report No. 2*, 50 Congress, 2 Session (1888-89), pp 33-37; *Mining & Scientific Press*, February 2, 1889, p. 75.
50. *Sentinel*, September 26, 1891; *Mining & Scientific Press*, June 13, 1891, p. 372; October 3, 1891, p. 45.

Slag Pots were used to draw off furnace impurities. The ordinary slag pot was a simple conical cast iron vessel, two feet high, with three legs upon which it stood, and three trunnions by which it was handled. It weighed 400 to 500 pounds, according to thickness. Heat of the slag ultimately cracked them, as in Figure I. To make the pots more durable, they were later fabricated by Hawkins & Cantrell, San Francisco foundrymen, in replaceable sections as illustrated above. (*Mining and Scientific Press,* April 26, 1873)

CHAPTER FOUR

SALT LAKE VALLEY, METALLURGICAL CENTER OF THE WEST

In point of time Utah was the next place to develop a smelting industry, although lead ore from a locality 140 miles south of Salt Lake City had been reduced as early as 1858. That enterprise was not motivated by commercial desire, but stemmed from an 1856 mandate by Brigham Young which allowed his Mormon followers to search for deposits of lead for home use, Utah then being under threat of invasion by a United States army. Responding to reports of occurrences of lead near Beaver, Bishop Tarlton Lewis searched the area in 1856 without success. The following year he tried again in company with an Indian who led him to a "dry, light, iron-stained lime rock," a formation known as "dry bone," characteristic of cropping of lead mineralization. A little excavation by Lewis uncovered solid galena or lead sulphide.

The next year (1858) a company was formed consisting of Bishop Lewis, Isaac Grundy, James H. Rollins, William Hyde, and several others. They located other lead deposits and erected a draft furnace. A mining district was organized, called the Pioneer, and James Rollins elected its recorder, a position he would hold until at least 1875. The town of Minersville sprouted near the claims. During fall of 1858 and spring of 1859 three and a half tons of lead bullion were refined from ore taken from what would become the Rollins Mine. Cast into small ingots, the lead was sold for 25 cents per pound throughout southern Utah. In winter of 1858-59, two weeks before arrival in Utah of General Albert Sidney Johnston's army, 650 pounds of lead was taken from the mine and sold for 20 cents per pound. This metal was cast into bullets. Fortunately, the bullets were not used, for Johnson's Army marched through Salt Lake City and occupied Camp Floyd, southwest of the Mormon capitol.

Over course of the next several years other furnaces were erected, causing production to soar. By 1864 fourteen tons of lead had been processed from the Rollins Mine, some of which was transported as far as Laramie where it was sold for local consumption. Fending off Federal intervention into their affairs, and preoccupied in scratching a livelihood out of Utah's soil, the Mormons never considered the precious metal content of the lead. That would be left to gentile invaders.[1]

The Civil War and national policy growing out of that conflict ultimately led to commercial exploitation of Utah's mineral resources. At the first sign of succession, General Albert Sidney Johnston departed Utah to put himself at the disposal of Jefferson Davis and the Confederacy. With opening of hostilities the entire grand army of the Utah Expedition, sent westward in 1857 to resolve the "Mormon Question," evacuated Camp Floyd and marched eastward. For the moment Utah was isolated and the overland route to California left unprotected. That situation was remedied by the War Department in July 1861 when the Second and Third California Volunteers, units composed largely of miners, were activated to guard the overland mail. A tough, uncompromising, 41-year-old Irishman commanded the contingent.

Colonel Patrick Connor — his surname was actually O'Connor on his original enlistment papers — emigrated to the United States at an early age with his parents from County Kearney, Ireland; and his military career began with the First Dragoons on November 28, 1839, at age 19. Upon conclusion of five years of service he signed up with the Texas Volunteers on May 6, 1846, using the name Connor, by which he was known thereafter. During the Mexican War he was wounded at the Battle of Buena Vista and discharged with the rank of captain; shortly thereafter he joined the gold rush to California. Connor fared well on the Pacific Coast, engaging in the lumber business at Stockton. Although a respected businessman and public servant, Connor maintained his ties with the military by holding various positions with volunteer units. As a member of the California Rangers, he pursued the bandit Joaquin Murietta in May 1852. In the chase the desperado suffered nearly a dozen wounds before being fatally shot by Connor.

When orders were issued in July 1861 to activate military protection for the overland mail, Connor responded to the call and on September 3, he signed up as colonel of the California Infantry. On

July 5, 1862, the War department directed the unit to Utah, and a week later Connor, as commanding officer of both the Second and Third California Volunteers (an aggregate of 1,000 infantry, 500 cavalry, a battery of field artillary, and 200 wagons), started the long trek into the Great Basin.

News that Connor did not intend to reoccupy Camp Floyd, but would instead cross the Jordan River and establish a military post on the benchland immediately east of Salt Lake City, raised Mormon ire. Believing that surveillance was the true motive for the army's presence, Brigham Young seriously considered forcibly resisting Connor's crossing of the Jordan. But Connor was "a blessed hard man to scare," as the battlefields of Mexico testified, and he sent word to Mormon authorities that he "would cross the river Jordan if hell yawned below him." With loaded rifles, fixed bayonets, and shotted cannon, the advance guard of Connor's command marched into Salt Lake City on October 21, 1862. There were "neither cheers nor jeers" from persons lining the streets, remarked one observer. The command halted briefly before the mansion of Governor Stephen S. Harding, a gentile who had little love for the Saints, to hear a bombastic speech and then proceeded on its march through Salt Lake City. Upon the slope between Emigration and Red Butte canyons the column halted and went to work constructing what was to be an endless irritant to Mormons — Camp Douglas.

It was no easy task to bring an armed contingent into Utah Territory among Mormon pioneers who had been widely persecuted, who came to a desert wilderness no one wanted, and who felt secure in the haven they had wrested from nature. Through diplomacy, coupled with understanding, sympathy, and a belief in the inexorable march of progress, Connor was able to prevent hostilities between his forces and the saints. The colonel's immediate task was to protect the overland mail route; his ultimate objective, however, was to reconstruct Utah and neutralize Mormonism, which he considered "not only subversive of morals, in conflict with the civilization of the present age, and oppressive on the people, but also deeply and boldly in contravention of the laws and best interests of the nation."

As Connor saw it, there were only two ways to strike at "the baneful influence" of Mormonism. The first, military force and martial law, he considered impractical; and the second and by far the wisest course, was to invite into the territory large numbers of gentiles. Only discovery of gold, silver, and other valuable minerals could "induce an immigration . . . of a hardy, industrious, and enterprising population." Such an immigration would once and for all settle the "Mormon Question."

Connor inaugurated his policy by questioning Indians, traders, army employees, Mormons and apostate Mormons, as to mineral occurrences. He instructed his officers to furlough their men to prospect the country "whenever such course would not interfere with their military duties." The results, he wrote to his commander-in-chief, "have exceeded my most sanguine expectations." Metalliferous deposits were located in the ranges in the vicinity of Salt Lake City; on September 17, 1863, the first mineral location in Utah (the West Jordan claim) was made in Bingham Canyon on the eastern slope of the Oquirrh Mountains, and the first mining district (the West Jordan) was formed. Discoveries were rapidly made thereafter. Outcrops of argentiferous galena were located in Little Cottonwood Canyon in 1864, only to be abandoned because of the great expense of working them. In 1867 most of these claims were jumped and the Mountain Lake District formed, which included most of the Wasatch range. In 1869-70 this district was subdivided into the Little Cottonwood, American Fork, and Uinta districts.

Strikes of silver-bearing ore in East and Dry Canyons on the east side of the Oquirrh Mountains were made by soldiers of Connor's command then stationed at Camp Douglas. In March 1864 the town of Stockton on the west side of the Oquirrh Mountains was surveyed, organized, and made a military post known as Camp Relief, for the purpose of herding stock and obtaining a supply of hay. That reservation included the area about Rush Lake, then but a small pond. The following month soldiers detailed for duty there explored nearby mountains and discovered silver-lead outcrops, and organized the Rush Valley Mining District, which embraced all of Utah west of the Oquirrh Mountains.

Soon after the organization of the district, Lieutenant James Finnerty erected a small furnace to test the ore, and a larger smelter was built by the Rush Valley Furnace & Smelting Company. Both were only partially successful owing to the imperfect quality of fire brick used in construction. Subsequently Finnerty built a second furnace and ran it with good results for a number of weeks, turning surface ore into bullion. Ignorance of metallurgy hindered separation of lead from the precious metal, and Finnerty failed. Others stepped forward to try their luck.

In 1866 two companies erected furnaces. About a thousand pounds of bullion, assaying $228 in silver to the ton, was produced. Although successful, operating costs worked against these smelters. Supplies and labor were high, and cost of transportation — $250 to $300 per ton to the Pacific Coast and $300 to $400 to the Atlantic — ate up any profit in marketing either ore or bullion. Such was the case at close of the Civil War. Regular soldiers from the East relieved the volunteers at Camp Douglas. The California and Nevada soldier-miners returned home, the Mormon Church boycotted all things gentile, and mining languished in Utah for the next several years.[2]

In 1869 the Central Pacific and Union Pacific railways were connected at Promontory, Utah, making an uninterrupted transcontinental line. This prompted the heirarchy of the Mormon Church to enter the railroad-building act. In spring of 1869 Brigham Young organized the Utah Central Railroad. Ground was broken at Ogden in May and track-laying began on September 23, 1869. By January 10, 1870, this standard gauge road was complete to Salt Lake City, a distance of 37 miles. The cost was $1,500,000.

As a business enterprise the Utah Central was a remarkable success; its construction a noteworthy event in the history of silver-lead mining west of the Rocky Mountains. Once completed and linked to the transcontinental railway, the line furnished improved transportation facilities in Salt Lake Valley. Goods from Eastern manufacturers could easily be forewarded to any point in Utah. Likewise, Utah products — livestock, farm produce, lumber, coal, charcoal, ore and bullion — could be sent to markets east and west. Although a great boon was at hand in 1869, mining did not advance much in Utah because freight rates did not immediately drop.

Within a year other factors combined with improved transportation to change the direction of Utah mining. Argentiferous lead ores were successfully treated on a commercial scale at White Pine and Eureka, Nevada. A large mining population close at hand demanding Utah products suddenly nullified Mormon opposition to gentile enterprises. And in 1869 the Emma Mine in Little Cottonwood Canyon was discovered. In the first year of its operation, the Emma Mine became a big producer; with a total shipment for 1870 of 5,293 tons of ore with an aggregate value of $967,000.

The opening of that bonanza, "a lake of mineral of vast extent," which yielded a profit of nearly $120 per ton on shipments of ore to

Swansea, Wales, "gave an impetus to mining in Utah that surpassed all other efforts made in that direction put together." Prospectors, technicians, metallurgists, engineers, and entrepreneurs — men who had known of Utah's mineral wealth since Civil War days — descended upon the territory. Mountain slopes and canyons were combed and lodes of silver-bearing galena located adjacent to the Emma Mine in Little Cottonwood Canyon. Other deposits of argentiferous lead were found in Big Cottonwood, American Fork and Bingham canyons; at Stockton, in East Canyon (respectively 40 and 55 miles southwest of Salt Lake City), and at several places in Tintic Valley, 70 miles southwest of Salt Lake City.

Miners saw little difference between the ores of Utah and those at White Pine and Eureka, Nevada, and believed that it was more expedient to smelt ore near the mines and ship the bullion to refineries at San Francisco, Newark, Chicago, Omaha, or St. Louis. The immense amount of ore streaming from mines in Little Cottonwood Canyon produced a mania of furnace construction. In 1870 Francis P. Pascoe erected a furnace a quarter of a mile north of Warm Springs Bath House. And there was Robbin's Works, a large reverberatory roasting furnace and a shaft furnace, located about eight miles from Salt Lake City on the road to Cottonwood. Although Robbin did not get started until August 1871, owing to difficulty in obtaining a regular supply of sorted ore, his smelter was the only one at that time using Green River coal as fuel. The shaft furnace, however, was fueled with charcoal.[3]

Most furnaces, however, were built at or near the mines. In late 1870 I. C. Bateman and David E. Buel, fresh from their success at Eureka, Nevada, arrived in Utah with plans of the latest furnace design. Joining with Ellsworth Daggett, a knowledgeable metallurgist from New Haven, Connecticut, they built a battery of Piltz furnaces in Bingham Canyon to treat galena from the Winnamuck Mine. Known as the Winnamuck smelter, these furnaces were incorporated into the Utah Smelting Works.[4]

At Granite, near the mouth of Little Cottonwood Canyon, Buel and Bateman a few months later erected two Piltz furnaces. Three feet, six inches in diameter, with five tuyeres, these smelters were built of imported fire brick and lined with fire clay from Camp Floyd, which answered "the purpose required very well." Both works of Buel and Bateman were identical to the first Piltz furnaces

FIG. 1.

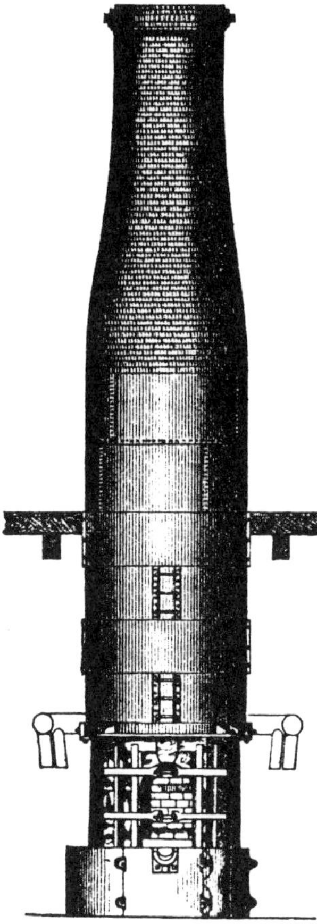

FIG. 3.

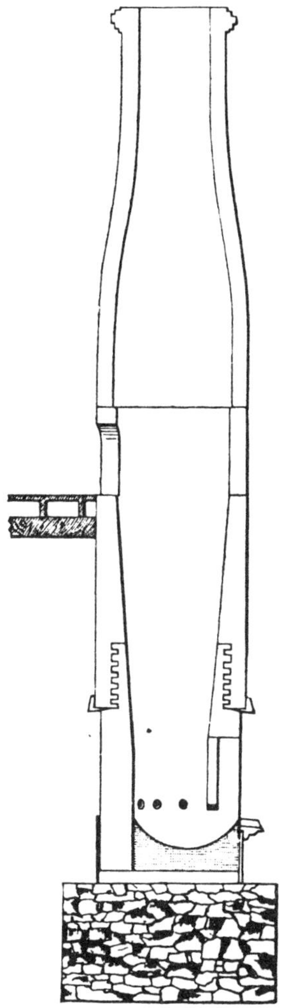

FIG. 2.

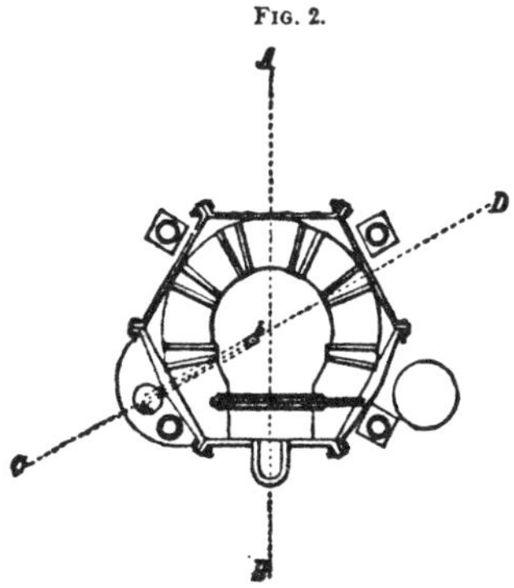

FIG. 4.

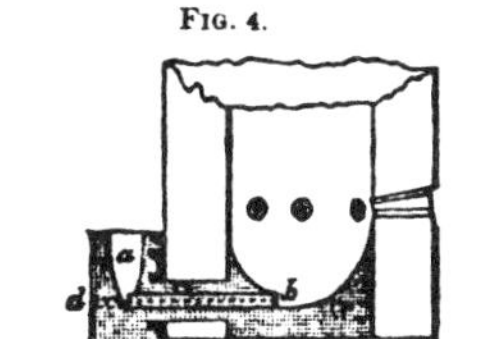

Diagram of the Piltz furnace installed at the Winnamuck smelter in Bingham Canyon. Figure 1 is a front elevation of the furnace showing the columns and their attachment to the flange; Figure 2, a horizontal section through the tuyeres; Figure 3, a vertical section through the line A B in Figure 2; and Figure 4, a vertical section though the line C D in Figure 2, showing the lead well and siphon-tap. (From Daggett, "Smelting in Utah")

The Winnamuck Smelter, Bingham Canyon. This smelter started out with one furnace, turning out about five tons of bullion a day. In early April 1872 a second furnace was added, and by June the Winnamuck was considered "the crowning achievement of mining enterprises in Utah." (From *Utah Mining Gazette*, June 27, 1874)

erected in Nevada by Karl von Liebenau at the Eureka Consolidated and Richmond companies.

Design of the furnaces were "simple," reported the *Salt Lake Turbine* on July 14, 1871, "the fireplaces easily and quickly repaired when burnt out." In other words the furnaces had removal hearths. A fine water privilege and a turbine wheel powered Roots blowers. Truckee charcoal was the fuel, at a cost of 27 cents per bushel. Buel and Bateman got off to a shaky start in early 1871, the snows holding up ore shipment, but by spring they were in "regular production," running night and day on ore from the Flagstaff Mine, producing six tons of bullion daily.[5]

Also situated at the mouth of Little Cottonwood Canyon, about three-fourth of a mile south of Buel and Bateman's furnaces, were the works of the Davenport Company. Erected between August and November 1872 by Jay H. Wooley, "a well-known furnace and mill builder," the works consisted of two vertical blast furnaces using water as motive power for the blast. The smelter processed ore from the Davenport Mine, using charcoal furnished by Gordon & Murray. Four years later the furnace passed into the hands of Mather and Geist.[6]

At a point called Tannerville, six miles up Little Cottonwood Canyon, were the furnaces of Jones and Pardue. These proprietors were fortunate in that they were close to groves of quaking aspen and spruce pine, which they cut and burned into charcoal to fuel their furnaces, which commenced operation in early July 1871. Jones and Pardue, however, sold their furnaces to Wellington Mining and Smelting, a company incorporated in San Francisco in January 1872. The shortlived Moniter furnace, built in 1871 by T. R. Scheuner, was also situated at the mouth of Little Cottonwood Canyon, at a spot known as Union Fort. It consisted of a blast furnace having a daily capacity of 10 to 15 tons.[7]

While all this was going on, argentiferous galena was discovered 12 miles up American Fork Canyon. A consortium of wealthy men, consisting of the Vanderbilts of New York, Lloyd Aspinwall who built a narrow gauge railroad across the Isthmus of Panama, a Mr. Howland of New York, L. A. Hardins of Chicago, and R. N. Baskin of Salt Lake City, purchased the principal claims and incorporated the Miller Mining Company. In an amphitheatre within a few miles of the divide between Little Cottonwood and American Fork canyons, the company built the Sultana Smelting Works in spring of 1871.

In several respects the Sultana Smelting Works was a revolutionary refinery. First, it had state-of-the-art furnaces, secondly, capable managers. The works consisted of three vertical Piltz furnaces, and a roasting oven. A two and three-quarter mile long mule tramway connected smelter and the mine. Thanks to the company's general manager, Major Edmund Wilkes, the company possessed one advantage over other firms engaged in the manufacture of bullion. It took steps to produce its own fuel. Close to the smelter a sawmill was erected to furnish cord wood to power the steam engine which

The Sultana Smelter, American Fork Canyon, showing only one stack. (*Utah Mining Gazette*, November 29, 1873)

drove the smelter's blowers and auxiliary machinery. Most important, the sawmill furnished cord wood to process into charcoal, the smelter's fuel.

Because the Miller Mine and the Sultana Smelter was at high altitude (7,000 feet) in the Wasatch Mountains, where snow impacted charcoal making, the fuel was not burned in heaps. The Miller Mining Company brought in James C. Cameron, Jr., a metallurgist experienced in charcoal-iron making in the Lake Superior region. He brought with him the concept of a beehive shaped masonry charcoal kiln, a structure not affected by inclement weather.

Under the supervision of Hugh McAllister, who had worked with Cameron at the National Iron Company in Wisconsin, 15 kilns were built on bench land a few hundred feet above the smelter. A tramway supported by trestles connected the kilns with the furnace fill floor. Thus charcoal, a fragile commodity, could be dumped at the mouth of the furnaces, eliminating waste by being handled only once.

With the sawmill going night and day, the kilns produced 1,500 bushels of charcoal per day, more than enough fuel to run three furnaces. Excess fuel was stored in sheds. Not only did the kilns halve the smelter's fuel costs, they provided additional revenue through the sale of excess charcoal to other smelters. More will be said about the importance of the beehive kiln.[8]

Close to the smelter, sawmill, and kilns grew Forest City, a town which ultimately attained a population of about 500. Making fair wages, its men supported two stores, two saloons, a billiard hall, and a boardinghouse. For a while at least, it was an "active little burg."[9]

* * *

By late 1870 two obstacles were blocking the path of Utah's smelting industry. The first and most serious impediment was that most of Utah's argentiferous lead ore was not self-fluxing. Hematite, an iron ore, had to be transported from Wyoming over the transcontinental railroad, and shipped by the Utah Central to Salt Lake City, and then freighted to canyon smelter sites. Transportation charges ate deeply into smelter profits. The second problem was accessibility of cheap fuel. With exception of the Sultana Works and Jones & Pardue, smelters had to import charcoal from Truckee, California, which was marginally successful. A fraction of the charcoal received in Salt Lake Valley was usable, the long railroad trip reducing much of the fragile commodity to unworkable powder. It soon became apparent to all mining men that a glaring logistical problem had to be corrected if Utah's budding mining industry was to survive and grow. Men of foresight believed there was only one remedy, a railroad system had to be built linking Utah's mining districts to the transcontinental railroad.

A grand trunk line, called the Utah Southern Railroad, was the brainchild of Mormon leader Brigham Young. Organized in January 17, 1871, with Brigham as president and John Sharp superintendent, this standard gauge line, capitalized at $1,500,000, was intended to link the northern and southern portions of Utah Territory. Payson in Utah County, 75 miles south of Salt Lake City, was fixed upon as terminus; but the road was envisioned extending south to the headwaters of the Colorado River and eventual connection with the Southern Pacific Railroad.

Ground was broken on May 1, 1871, at Salt Lake City's Utah Central depot, and 40 miles of track laid to Lehi in Utah County. By January 1875, the road had been extended to Santaquin, 20 miles south of Provo. Although this does not sound like much of a railroad, the Utah Southern was of inestimable value in the development of the territory's mineral resources.[10]

In its southward run from Salt Lake City, the Utah Southern passed through Jordan Valley. Situated at a strategic spot in the valley, 12 miles south of Salt Lake City, was the agricultural community of Sandy. Because Sandy was midway between Cottonwood and Bingham canyons (17 miles from either place) it became the distribution center for supplies for both mining districts. At the station, all supplies for both camps were transferred from rail cars to wagons, and Sandy received the products of both Bingham and Little Cottonwood canyons, ore and bullion. Sandy Station, as the town came to be called, also had another thing going for it, it was close to the Jordan River.[11]

Sandy's location on the Utah Southern Railroad, facilitating easy connection to the Utah Central Railroad, and its proximity to water were factors no financier could overlook. Bankers, merchants, and mining men pooled their resources, and within a year Sandy Station was transformed from a sleepy agricultural locality to a busy transportation hub, the receiving point for the wealth of the Wasatch and Oquirrh mountains.

The first to enter the act were Lloyd Aspinwall and Edmund Wilkes, owners of the Miller Mining Company and the Sultana Smelter in American Fork Canyon. In April 1972 they incorporated the American Fork Railroad, a three foot narrow gauge line, to provide a rail connection to the Utah Southern at American Fork City, three miles south of Lehi. Capital stock was $300,000. Rail laying commenced on May 20 at Provo, a station on the Utah Southern line, near Lehi. By October 10th 16 miles of road had been laid, the first six miles approaching the canyon being the easiest to construct.

Laying track up American Fork Canyon was a different story. The defile was narrow, tortuous, with grades reaching 297 feet to the mile in some places. The river had to be bridged 16 times in the first five miles. By end of 1872 rails had reached Deer Creek, 16 miles from the line's connection with the Utah Southern. An additional four miles was graded to the Sultana Smelter, but track was never laid, the numerous switchbacks and grades being beyond the capacity of the line's Mason Bogie locomotive. A hotel was built at Deer Creek in 1877 to accommodate passengers.[12]

Hugh White, R. N. Baskin, and the Walker Brothers, the latter Salt Lake City merchants, were the next to launch a railroad building scheme. In September 1872 they pledged $300,000 worth of capital stock, and incorporated the Bingham Canyon and Camp Floyd Railroad, a narrow gauge line that would run from Sandy Station to Bingham, thence to Tooele City, then to Stockton and Ophir and on to Camp Floyd. A branch standard gauge line (four feet, eight inches), with a third rail for narrow gauge trains, would be constructed from Sandy to the Jordan River, where the largest and most extensive smelters of the Territory were being built.[13]

Grading from Sandy Station began in late 1872, and by March 1873 ties, iron, and engines were on the ground and track laying began in earnest. Although active operation from Sandy Station to Bingham City was scheduled to begin April 1, 1873, passenger service did not start until October 16; reason for the delay being that White's resources were exhausted.

Out of money, White sold control of the railroad to a syndicate headed by New Yorker Charles W. Scofield, who had mining interests at Bingham. Remaining with the railroad, White superintended its completion on November 23, 1873, to the Winnamuck smelter and mine in Bingham Canyon. A stage took passengers the last five miles, delivering them to Bingham City by 9:30 in the morning, allowing "them ample time for the transaction of business in that thriving camp, and return again to Salt Lake the evening of the same day."[14]

Although the Bingham Canyon Railroad was never pushed over the Oquirrh Mountains to the Camp Floyd district, it benefited Bingham Canyon. At last supplies and mining machinery could be transported fast and cheaply to the mining camp, and ore and bullion sent down to Salt Lake Valley with equal facility. In six months the terminus of the Bingham Canyon Railroad near the Winnemuck become "quite a point for trade," sporting a large boardinghouse, a forwarding house and other buildings. The people at the terminus were sociable in their habits. "Scarcely a week passes without amusements of some kind or others," reported the *Tribune* on March 17, 1874.

The Bingham Canyon Railroad was an immediate success. In February 1874, 3,891,749 pounds of freight was forwarded from Bingham Junction, and 4,728,545 pounds were received, making a total of 8,620,294 pounds handled. Within a year the railroad was "one of the best paying properties in the Territory."[15]

While the Bingham Canyon Railroad was pushing into the Oquirrh Mountains, another narrow gauge line was creeping toward Little Cottonwood Canyon of the Wasatch Mountains. Organized October 24, 1872, and capitalized at $500,000, the Wasatch and Jordan Valley Railroad was the creation of its president, William Jennings, and Frank Fuller, its secretary. The company's principal stockholders were officers of the Emma Mining Company in Little Cottonwood Canyon, and members of the Mormon heirarchy who wanted to bring granite from a quarry at the mouth of Little Cottonwood Canyon for construction of the Salt Lake City temple. Planned to run from Sandy Station to Alta, the road would be double track from Sandy to the mouth of the canyon, one track being 4 feet 8½-inch gauge, the other three feet. The gauge through the canyon would three feet. Provision was made to protect the road from snow slides, the big hazard during winter.

Grading began from Sandy in December 1872, and it was contemplated to have the road in operation to Granite at the mouth of the Canyon 90 days later. It would then be pushed up the canyon as rapidly as possible. By April 1873 seven miles of the road were built from Sandy to Granite. Coaches having not arrived from the East, passengers were transported on flat cars, a pleasant experience so long as the weather was good. A *Tribune* correspondent wrote: " The trip was a very . . . exhilirating one, and the time passed in making the journey to and fro was fraught with much interest. . . . The scenery on the right towards Salt Lake Valley, as the road approaches Little Cottonwood Canyon, is grand and majestic in the extreme; there you have a full view of Camp Douglas, Utah Lake, the Great Salt Lake, Church Island and . . . other minor islands which dot the surface of that wonderful inland piece of water."[16]

A scenic ride, the Wasatch & Jordan Valley Railroad was a tough construction project with three to six percent grades up the canyon; the greatest being 212¼ feet to the mile. And there were numerous turns; in one spot the radius was 38 degrees, on approach to Granite it was eight degrees. By September 1873 the line had been pushed only eight miles to Wasatch in Fairfield Flat. With money running out, the line was sold in June 1875 to Charles Scofield, who linked the line to mines at Alta with a seven mile long mule tramway. On April 29, 1879, Scofield merged the line with the Bingham Canyon & Camp Floyd Railroad, the stronger of his two railroads.[17]

* * *

The inauguration of a narrow gauge railroad system was a boon to Utah's mining industry. It afforded easy transportation of ore, fuel, and flux to a centralized point where there was level land adjacent to a flowing stream. All the components were present at Sandy Station for large scale smelting operations. Within three years a half dozen smelting works had been erected, giving the station on the Utah Southern Railroad the appropriate sobriquet of the "Swansea of Utah."

The Saturn Silver Mining Company, Ltd., was the first to capitalize on the location of Sandy Station. In early 1872 this English-owned company, capitalized at £75,000, acquired land a quarter mile south of Sandy Station, on the Utah Southern Railroad. Under management of G. M. Gerrish, an old White Piner, three vertical blast furnaces were built, each with a daily capacity of 20 tons of ore. Blast was furnished by a six foot Sturtevant blower driven by a 30-horsepower steam engine. Commencing operation in May 1872, the works gradually increased production until it was making the largest runs of any smelter in Utah Territory. Under management of Francis Fowler, in a little over two weeks in fall 1872, this custom smelter produced 247 tons of bullion.[18]

The Saturn Works had its ups and downs, however. On September 15, 1872, the works were leased to Theodore F. Tracy of Salt Lake City, and John Gray served as his manager. Under their control, the works processed ore from mines in Little Cottonwood Canyon: the Emma, Vallejo, Last Chance, and Frederick. It also drew ore from Bingham Canyon mines: the Hidden Treasure, Spanish and Silver Hill. Small lots of ore from Big Cottonwood, Tintic, and adjacent districts, were also taken in. Using limestone from Warm Springs near Salt Lake City, hematite from Rawlins, Wyoming, and charcoal from Truckee, California, and Piedmont, Wyoming, the smelter turned out on average 65 tons of bullion per week, with $175 in silver per ton.

Extension of narrow gauge lines up the canyons produced a flurry of smelter building at Sandy in 1873. That spring the Last

Chance Mining Company of Bingham Canyon erected two furnaces, housed in a "fine, capacious building" a few hundred yards east of the station on the line of the Wasatch and Jordan Valley Railroad. Superintendent of the works was Samuel M. Wessels, formerly manager of the Saturn smelter.

The most extensive plant erected at Sandy was the Myers Smelting Works, also called the Galena Smelting Works or the Sandy Smelting Company. Owned by John W. Kerr, of Salt Lake City, and under supervision of Frederick H. Myers, this smelter was built on the east side of the Jordan River, at the crossing of the Bingham Canyon Railroad. A correspondent, writing in the April 29, 1873, issue of the *Tribune*, stated the smelter was well placed: on a plateau, "the top being some 150 feet higher than the base, thus affording facilities for construction of different floors and furnaces required for economic, labor-saving and expeditious manipulation." The smelter was water powered, Myers tapping the Jordan River three and half miles above the works. A canal conducted water to the works, and a double turbine waterwheel furnished power equivalent to 125 horsepower.

The Galena Smelter was a modest refinery at first, consisting of only one furnace capable of turning out eight tons of bullion daily. With a desire to abolish the use of charcoal, Myers fired up the furnace on April 18, 1873, using a mixture of bituminous coal, coke and wood, to process ore from the Galena Mine in Bingham Canyon.

In July 1873 the furnace passed into the hands of Carson and Buzzo with purchase of the Jordan and Galena mines. They constructed three more furnaces, and a fifth furnace was added in 1874. All five furnaces, called the "Old Works," were enclosed in one 160 by 35 foot building. A series of spur lines and switches conducted ore, flux, and fuel to bins enclosed in a 160 by 40 foot building. Another track was laid before the furnaces for shipment of bullion.[19]

As well engineered as the Galena Smelter was, it had weak points. Freezing weather often stopped the flow of water through the canal, shutting down the double turbine wheel. When that happened the works were compelled to rely on its steam engine, which was only able to furnish power to four out of five furnaces.[20]

In 1876-77 the Galena Smelter passed into the hands of a Captain Selfridge. He leased three of the furnaces to Liberty E. Holden, owner of the *Cleveland Plain Dealer*, who with his son, Albert E., had recently become interested in Utah mining properties. The

Holdens tore out the old stacks and replaced them with others of an improved pattern. They were fired up in spring of 1877, and ran on ore furnished from the Jordan Mine of Bingham Canyon, and on custom ores from the Cottonwoods. Having gotten ahead of the story, we need to backtrack to the period of smelter building at Sandy.[21]

Several other smelters were built at Sandy in 1873. The Sheridan Hill Mining and Smelting Company of Utah, composed of Israel, Isaac, and Emanuel Schoenburg, and Simon and J. R. Bamberger, was incorporated in New York on February 24, 1873, to work ore from the Grizzly and Kempton mines on Sheridan Hill in Bingham Canyon. Capitalized at $600,000, the company erected within 200 yards of the Galena Smelter a building with two blast furnaces, driven by a 40-horsepower oscillating engine. Because the machinery was manufacturerd by the Morgan Iron Works of New York City the smelter was nicknamed the Morgan. Among the improvements introduced was a novel application of water and wind boxes in connection with the furnaces, also a patented method of filling and emptying the bullion molds. The smelter was in operation by June 1873. Three additional furnaces and a 60-horsepower engine were erected by year's end, making a total of five furnaces. Average furnace production was from 160 to 180 bars of bullion per day.[22]

Two hundred yards south of the Sheridan Works, the Utah Silver Mining Company of Bingham Canyon also acquired land to erect a smelter. Working ore from the Neptune Mine, in 1874 this company produced 4,623,000 pounds of bullion, containing $16,610 gold; $297,285 silver, and $184,341 lead.[23]

Just south of the Saturn Works, the Mountain Chief Smelter, consisting of two furnaces, was constructed in spring 1873. Shortly thereafter this smelter's name was changed to that of Mingo. Idled in 1875-76, it was leased by the Pennsylvania Lead Company, a Pittsburgh firm, which built several additional furnaces. By 1877 the Mingo was proclaimed "one of the neatest and most convenient smelters in the country, of a capacity of 15 tons of bullion daily." Under supervision of A. K. Knapp, the Mingo Smelter ran entirely on custom ore.[24]

The Last Chance Furnaces were erected a quarter mile east of Sandy in 1873 to smelt ore from the Flagstaff Mine of Little Cottonwood Canyon, and the Last Chance Mine at Bingham. This smelter was owned by the Last Chance Mining Company, a London corporation which also owned the Flagstaff Silver Mining Company, Ltd., and the smelting works at the mouth of Little Cottonwood Canyon built by Buel and Bateman in 1871. Plagued by frequent shutdowns due to difficulty of getting ore down the canyon during winter, the Flagstaff Company leased the smelter to Mather and Geist, practical smelters of Salt Lake City, who spent $40,000 repairing the old works.[25] Freed of that burden, the Flagstaff Company poured money into its Sandy property, building three steam-driven blast furnaces. Two more furnaces were added in 1874, making it one of Utah's best smelters.[26]

According to the *Mining and Scientific Press,* in 1877 the Last Chance Furnaces were run in "first class style." Employing 90 men, it reduced 120 tons of ore every 24 hours, yielding about three car loads of base bullion daily, a greater quantity than produced at any other smelter in the territory. "Everything about the premises goes forward with the regularity of clock work," noted the *Mining and Scientific Press.* "Each man knows his duty and performs it. The discipline is regular, and strict compliance is exacted from all. A man who gets intoxicated is at once dismissed, and is never again employed, and like Squires, the management compels everybody engaged about the smelters to take a dose of medicine [potassium iodide, or even strychnia] before breakfast as a preventive against leading."[27]

Although not located at Sandy, the Germania Separating and Refining Works was the company which gave Sandy Station the title of "Swansea of Utah." Completed in November 1872 the works were situated on 25 acres on the Utah Southern Railroad at Big Cottonwood Station, later called Murray, six miles north of the Davenport Smelter, and about eight miles south of Salt Lake City.

In 1873 Utah was producing 100 tons of base bullion per day, which had to be transported to New Jersey, Omaha, Chicago, or San Francisco for separation of precious metal. The three to five month delay for separation cost smelters and miners dearly, and greatly retarded Utah's base metal industry — the Germania Works was designed to remedy that situation.

When news broke that the company would be a refinery and not a smelter it was thought to be "very wild speculation." In 1872 few people in the Great Basin had the knowledge to build, let alone run such a refinery. But reality won out, and the business was successful, thanks to a thorough understanding of refining techniques by

The Germania Separating and Refining Works. 1: blacksmith shop; 2, calciner; 3, offices; 4, assay offices; 5, three blast furnaces; 6, engines and boilers; 7, refining cupelas; 8, three reverberatory furnaces; 9, desilverizing pots and two calciners; 10, softening furnace. (From *Utah Mining Gazette*, February 21, 1874)

its managers Henry Sieger and Gustav Billing, two knowledgeable German metallurgists. By 1874 the Germania was regarded as "the most successful works of the kind in the United States."[28]

Erected at a cost of $58,000, including land, the Germania Refinery consisted of a main building or furnace house, refining house, assay office, engine house and all necessary offices and out-buildings, together with manager's and superintendent's dwellings. In the furnace building were five large cast iron pots, set in substantial brick work, each one over its furnace or fireplace, and two reverberatory softening furnaces. Connected with this building was an engine house, containing a 45-horsepower boiler. A little to the north of this was an inclined shaft furnace, while to the west of the works, at a distance of a few rods, was a substantial brick stack, which was connected with the furnaces by means of underground flues of strong masonry. A side track was laid from the main track to the works, by means of which all materials were brought to their respective destinations.

The Germania Refinery employed the Flach process to separate lead and silver; the *modus operandi* being as follows:

> About twenty-three or twenty-three and a half tons of base bullion, as received from the different smelting works is put into one of the largest of the pots, which are five in number, two of them having a capacity of twenty-five tons each, the other three being smaller; the whole number being set in the form of a wedge, the two largest forming the broad part, while the smallest pot forms the point. A sharp heat is kept up until the impure lead is thoroughly melted, and sufficiently hot to melt the zinc (which is the medium of separation) immediately on its addition, or at a temperature of about 411 degrees Celsius, the melting point of zinc. Zinc is employed for the separation of gold, silver and copper from lead, owing to its possessing a greater chemical affinity for these metals than that of lead. The zinc employed at this establishment is of two kinds: viz: commercial or good zinc, and dross zinc (a refuse matter from galvanic batteries) which contains about thirty percent iron. The former is obtained from Illinois at a cost of nine cents per pound, and the latter from New York at a cost of five cents per pound. When the base bullion is tolerably free from impurities and contains from 150 to 200 ounces of silver per ton, two and a fourth to two and three fourths per cent of zinc, or three to three and a half percent of dross zinc suffices for its complete desilverization.
>
> As soon as the metal has attained the requisite heat, from a half to one per cent of zinc, or three-fourths to one and a half per cent of zinc dross is added to the molten mass, which is well stirred by two men for half an hour, and then allowed to remain still for three hours. Immediately after stirring, the fire is withdrawn from under the pot, and the metal allowed to cool to a dull red heat. The zinc being thoroughly mixed with the lead, takes up the gold and copper, with the greater portion of the silver, and rises to the surface, forming a scum or cake which gradually covers the whole of the lead.
>
> The object of adding a portion and not the whole of the zinc at one time, is to avoid the caking of the zinc on the surface of the metallic bath, whereby it would not be in a state of division sufficiently fine to mix thoroughly with the lead.
>
> At the expiration of three hours this scum is skimmed off

and transferred to the pot next in advance. The whole operation, from the commencement to this point, occupies about three hours, but varies with the amount of the impurities in the lead. A second addition of zinc is now made in the proportion of a one half to three-fourths per cent of zinc, or one to one and a half per cent of zinc dross.

The dross or scum in the second or advanced pot, after being melted, stirred, and allowed to cool, is skimmed and the skimmings transferred to the third, or end pot of the series, the alloyed lead and zinc containing a little silver which has sperataed from the dross being moved back to No. 1. This is effected by means of an iron gutter leading from the rim of one pot to that of the other, into which the lead is ladled. A similar process is then performed with pot No. 3. The dross being skimmed off and transferred to a receptacle called the "safe," and the separated alloy of lead and zinc carried back. The remainder of the zinc having been added to pot No. 1, it is again stirred and skimmed, the skimmings being put aside for further operation.

The lead remaining in pot No. 1, is now practically free from gold, silver and copper, but is rendered a little hard by a portion of the zinc remaining in it, together with a little antimony, of which, if there is much in the base bullion, a small percentage remains.

At the bottom of the large pot is a pipe connected with an iron gutter leading to the hearth of a reverberatory softening furnace, which is placed at a lower level than that of the pots, to avoid the necessity of moving the lead by hand.

The furnaces of which there are two, one for each large pot, are fifteen feet six inches long and nine feet four inches wide, internal dimensions, with a hearth or bed of sufficient depth to contain the contents of the large pot. The bed of this furnace is composed of a large iron pan cast in three pieces and having the bottom flat and gradually curving to meet the upright sides and ends. This is covered with fire clay, and over the clay with fire brick set on end. When well put together these parts will last for years. The furnace has only one fire grate, placed at the end. These furnaces are built upon arches, which serve to keep the bottom cool, and in order to more easily remove any lead which through bad setting may trickle through. The furnace having been raised to the requisite temperature, the lead pot is tapped and the metal allowed to flow into the furnace. The heat is then kept up and a portion of the zinc and antimony is oxydized and escapes in the form of fume, which the remainder forms a scum which rises to the surface and covers the bath of lead. The lead is well rabbled from time to time with iron rakes, in order to expose a fresh surface to the air, and this is continued until the lead shows, by samples taken from the furnace, that it is pure and ready for tapping.

At a distance of a few feet from the end of the furnace is another pot, called the market pot, the level of its rim being a little below that of the hearth of the furnace, and into this the lead, after being skimmed, is allowed to flow through an iron gutter.

The moulds, which are on wheels, having been run under this pot, the lead is admitted into them by means of a tap which stops the supply as each mould is filled.

These bars of lead are made to weigh about one hundred and forty pounds each; and, when cool, are taken out and ready for shipment. They are now wheeled up an incline, but it is intended to have a raised tramway over the tapping place, onto which small wagons containing the bars will be hoisted, and run direct to the side track. This lead is now pure, containing, on an average, not more than half a pennyweight of silver to the ton, and being free from gold, copper and all impurities, and fit for any of the purposes to which lead is applied, either for making white or red lead, sheet or pipe.

The whole operation, from the loading of the first pot to the tapping of the market lead, occupies twenty-four hours, and at the present time the capacity of these works is forty tons per day. . . .

The dross obtained from the last skimming of No. 1 pot is transferred to a flat-bedded calcining or roasting furnace, of which there are two in the main building, one for treating the pot skimmings, the other for the skimmings from the reverberatory furnace. It is here thoroughly roasted at a temperature insufficient to soft it, so as to get rid of any remaining zinc or antimony, and when sufficiently roasted is then put in the flowing furnace, where the lead is separated from the iron and other matters which form a slag which is worthless. The level is tapped and returned to No. 1 pot, with a fresh charge of bullion, when the same routine is again gone through. The flowing furnace is very similar to the reverberatory softening furnace, differing only in the form and material of the furnace bed. The calciners, before mentioned, have each two grates, with a flue rising from the center of the furnace. The object of the two grates is to obtain a much higher degree of heat when necessary for the roasting of refractory or very impure skimmings.

The same process is gone through with the skimmings

from the softening furnace. The next point is the treatment of the rich alloy from pot No. 3, which was placed in the "safe." This is taken to the shaft furnace before mentioned, and which will be now described.

It is an inclined shaft furnace with the back sloping towards the front, at a little distance above the tuyeres the front recedes more abruptly from the back. The internal dimensions are, from breast to back two feet seven inches, and from side to side two feet six inches. It has three tuyeres having a diameter of only one and a fourth inches, and a pressure of blast equal to about twenty-four inches of water, or about one pound upon the square inch. The blast is produced by a blowing machine, twenty-four inches in diameter, and having a twenty-four inch stroke. The blast passes into a chest or reservoir, and is thence conducted by pipes to the furnace.

The fuel used is coke (that in all the other furnaces, with the exception of the crucible furnace, being Vandyke coal) which is obtained from Pittsburgh, and costs, delivered at the works, about $28 per ton.

The fluxes used are hematite iron ore, which is obtained from Rawlings, and costs about fifteen dollars per ton, delivered, and a little lead slag, of which plenty can be had in the neighborhood; the charge is composed as follows: rich alloy, 250 pounds; iron ore, 180 pounds; coke, 55 pounds; and a little lead slag. After a good heat has been raised in the furnace the charging is commenced; the coke is first thrown in and spread evenly over the surface. The rich alloy is then charged, being all fed against the front of the furnace, the iron ore and lead slag are then thrown in and mixed tgether, and spread evenly over the furnace, three of these charges are worked in one hour, so that about sixteen thousand pounds of alloy are passed through this furnace in twenty-four hours. The heat is maintained so low that the furnace at the feed door is perfectly black and quite cool; the only flame visible being a small reducing flame running up the inside of the front of the furnace. Owing to the peculiar construction of the furnace, the charge gradually sinks and meets the sudden incline, from whence the alloy passes down to the zone of fusion, in the center of the furnace, without touching the walls, and is not melted until when within a few inches of the tuyeres. By this process the copper is obtained as matte, and the gold, silver and lead, in the form of a highly enriched lead. This rich alloy is tapped off about once in every hour and a quarter, and the slag about every five minutes. This slag is received in cast iron pots where it is allowed to solidify before being turned out. The slag is almost entirely free from lead, and contains only a trace of silver, so that practically it may be considered clean. The bars of silver are now taken to the cupelling furnace. This is in appearance not unlike a very short reverberatory furnace with a very large grate; but instead of the usual bed are a couple of bars, on which is supported the cupellor, as it is usually called, the "test."

The test consists of an elliptical frame of wrought iron, filled in with bone ash well beaten, and hollowed out to the requisite shape by means of special tools for that purpose. At the back of the furnace are two holes, one where the blast pipe enters, and the other for introducing the rich bars into the test. The test, which is placed with its long axis towards the blast, is left a little thicker at the opposite end where the litharge flows over. When the test has been raised to a good heat, a sufficient quantity of the rich lead is introduced, and the whole speedily melts, and the blast is now on. The oxygen of the air composing the blast oxydizes the lead into litharge, and this takes up all foreign metals. A portion of the litharge is absorbed by the test, but much the larger quantity flows over the end of the test furthest from the blast, and is received in iron pots mounted on wheels, which are moved as fast as filled. The rich lead is continually added until all has been oxydized, and the silver is now pure, containing only the gold in the bullion. If, however, there should happen to have been more copper in the enriched alloy than could be carried off by the lead contained in it, (which is shown by the appearance of greenish spots floating on the silver) this can be easily removed by the addition of a little pure lead to the test.

Immediately on the termination of the cupellation a very beautiful phenomenon is seen, which is termed the "brightening." Just as the last trace of lead clears from the silver a peculiar bright and vivid flash seems to cover the whole bath of metallic silver, and for an instant every brick in the arch of the furnace is reflected as plainly as in a mirror.

A series of ingot moulds set in a frame are then run under the test, and a hole bored into the bottom of the test, through which all the silver runs into the moulds. These ingots are now taken to the crucible house, and about thirty-five pounds weight put into a plumbago crucible and melted. A little charcoal is thrown into the crucible, and the whole stirred with a stick. After skimming off the coal and any little slag which may have formed, the metal is poured into ingot moulds of the requisite size, and the silver, which is now fine, is after assay stamped with its degree of fineness.

This finishes the whole operation, as the gold and silver are not separated at these works, but sold as an alloy.

> Owing to the fact that all these operations are not proceeding at one and the same time, the number of men employed is comparatively small (being only thirty-five) when the large amount of work done comes to be considered.[29]

Although the Germania Refinery was the creation of Henry Sieger and Gustav Billing, in late 1875 a brilliant but authoritarian German named Frederic Anton Eilers, joined the business. Born in Nassau, Germany, January 14, 1839, he received his technical education at the University of Gottingen and the mining school of Clausthal. Coming to the United States in 1859, Eilers became an assistant in the New York City office of Adelberg & Raymond. From 1866 to 1869 he had charge of the Betty Baker copper mine and furnace in Carroll County, Virginia. When Dr. Rossiter W. Raymond became U.S. Commissioner of Mining Statistics, Eilers was his deputy, reporting between 1869 and 1875 upon districts where American silver-lead smelting was inaugurated.

Duty with the Bureau of Mining Statistics was an eye-opening experience for Eilers. What he observed at Cerro Gordo, Eureka, and other base-metal localities must have gone against the grain of this straight-laced, methodical German, whose metallurgical skill had been honed at Clausthal, where smelting was done according to strict Prussian rules. The hit or miss American metallurgical procedures baffled Eilers. Everywhere ore was shoveled into furnaces with the hope that it would make its own slag. Sometimes it would, sometimes it wouldn't. Freeze-ups and the digging out of sows were everyday occurrences, and at best the ratio of ore to fluxes was such as to be destructive of profits.

While American smelting was chiefly a rule of thumb affair, Eilers knew the vast mineral resources west of the Rocky Mountains would propel smelting into a great industry, and that Salt Lake City would become an important metallurgical center. With that in mind, in 1875 Eilers took over as general manager of the Germania Refinery. With characteristic thoroughness, Eiler went to work ironing out furnace irregularities that were then the rule rather than the exception.

Indeed, it was not difficult for a graduate of Clausthal to improve the Germania works. Eiler suspended desilverization at the refinery in 1875, enlarged the company's refining capibilities with the addition of six Du Faure tilting furnaces for distillation of zinc-silver-alloy. Three years later the company was again desilverizing bullion using the Parkes process. Eilers also broadened the company's financial base by installing three Raschette rectangular blast furnaces of 25 ton daily capacity, as well as a white lead department.

Eilers ran the Germania Refinery like a European barony, the plant and its employees running like clockwork. Just like the smelting works at Clausthal, everything at the Germania Refinery was regimented. The staff lived at the works and met daily at the mess with the superintendent at the head of the table and the juniors in their order. In that way the refinery, and Eilers, was a great training school for young metallurgists.

During his years at the Germania Refinery Eilers became a leader in furnace design and smelter construction. He designed a sectional water-jacket of cast iron with a flaring lip, called the "Eilers jacket." He also adapted to the lead smelting furnace the water-cooled tap-jacket, previously used in iron smelting, and thus replaced the clay breast of the furnace, assuring an almost continuous campaign. And he was one of the first, if not the first, to increase smelting capacity without increasing furnace width at the tuyere level.

Eilers' ability to keep the Germania furnaces running without interruption was a revelation not only to metallurgical pioneers of Salt Lake Valley, but to American smelters. It was the beginning of a professional reputation and business success which brought to Frederic Anton Eiler universal recognition as the "father of the modern practice of silver-lead smelting," and made him a "captain of industry."[30]

* * *

Sandy Station was a busy railroad hub in summer of 1873. Every train of both the Bingham Canyon and Wasatch & Jordan Valley railroads brought large quantities of ore from the mines of Bingham and Big and Little Cottonwood canyons. And bullion, the product of the smelters, piled up at the station at a rate of 150 bars per day, awaiting shipment to buyers in San Francisco, the Midwest, and East Coast.[31] This commerce transformed the rail station. A half dozen smelters decreed population, the needs of whom had to be met. Along with family residence came boardinghouses, stores, restaurants, bars, and billiard halls. The smelters were serviced by blacksmiths, pipefitters, tinsmiths, and sampling works, the latter an important adjunct to the smelting industry.

Sampling works came into being as a means of equitable arrangement between sellers and buyers of ore. Prior to their establishment, every dealer in mineral commodity believed their assays were the correct ones. As a result there was distrust, arguments, and delays. The samplers eliminated complaint by offering accurate assays to determine the value of large lots of ore. They also bought ore from miners, determined its composition and value, crushed and concentrated it, and resold the concentrates to custom smelters.

The first public sampling works in Utah was started by George J. Johnson, an ore purchaser, in the California Corral on Second South Street, Salt Lake City. Machinery not being available at the time, Johnson sampled by hand. Ore in lots of five tons and upwards was brought to the corral and sampled, purchase or refusal being the option of the purchaser. In June 1871 Johnson enlarged his works, calling it the Salt Lake City Sampling Works. Acquiring acreage at the corner of Fourth South and Third West streets, on the west side of the Utah Southern Railroad tracks, he erected a spacious building and installed a Brodie crusher and a French burr stone mill.

About the same time, in June 1871, H. S. Jacobs & Company established the Utah Sampling Works one block north of the Salt Lake City Works, on the eastern side of the railroad. The company's machinery consisted of a finely finished stationary engine and a Dodge crusher. During summer of 1872 the company passed into the hands of Baxter & Company, who converted it into an amalgamation plant. A year or so later the works was acquired by a Philadelphia company who leased it to J. C. Conklin who carried on the sampling business.

A number of sampling works were also established at Sandy between summer 1873 and spring 1874. The first to appear was the Pioneer Works, erected by Richard Macintosh on the line of the Utah Southern Railroad. He erected a large building (100 x 60 feet) and a sampling floor, and installed Howland and Brodie crushers, a Burr mill for pulverizing and a 25 square foot drying kiln. Platforms around the building accommodated a thousand tons of ore. The works had a capacity for sampling 250 tons of ore daily.

Also at Sandy was J. R. Nichols Metallurgical Works. Built on ten acres adjacent to the Utah Central line, it was the largest and best equipped sampler in the Great Basin. Nichols purchased for a period of 14 years exclusive right to use the Krom process in the Territory of Utah. This process, devised by Stephen R. Krom, a machinery manufacturer, was a dry mechanical process based upon the specific gravity

J. R. Nichols Metallurgial Works at Sandy.
(*Utah Mining Gazette*, May 16, 1874)

of ore. To house Krom's crushing and concentrating machinery, Nichols erected a three story 60 x 80 foot building with a basement. The structure, designed by Krom, cost $125,000 and consumed a half million bricks and 260,000 feet of lumber. Capacity of the works was 50 tons of crude ore per 24 hour day. The operation of Nichols Sampling Works was typical of samplers throughout the Great Basin.

Buying from mines or acting as their agent, Nichols took in great quantities of ore. When it arrived at the works, it was weighed in the wagon on scales occupying an attached building. The ore was then thrown into bins or piled in the open ward, every tenth shovelful as a rule being put into a wheelbarrow as a sample. That sample was spread out on the sampling floor (sand ore was worked up directly to obtain a thorough mixture). Hard ores was then passed through Cornish rolls. When thoroughly mixed, the ore was repeatedly quartered, dried, and further crushed, mixed, and quartered again. A portion was then ground to a consistency which allowed passage through a sieve of 70 to 80 meshes to the linear inch. The sample was then divided into three portions, one of which was assayed at the smelter, and the second at the mine or by a public assayer employed by the mine. If the results of the two assays agreed closely, a mean was taken as the true value of the load; otherwise the third portion was sent for control to an independent assayer.

The price smelters paid for ore was determined by a number of factors. From the assay value of silver in the ore a certain percentage was deducted for loss in smelting, which varied according to the nature of the ore, whether siliceous, ferruginous, or sulphureted, or according to special arrangement made between the sampler, mine owners, and the smelter. A further variable charge was made for cost of treatment, which was dependent on nature of ore and its tenor in lead. As a general rule in regard to oxidized ores, the charge for treatment was lower the larger the percentage of lead they contained. When the tenor was between 5 and 30 percent the lead was paid for at from 15 to 45 cents a unit of 20 pounds; the higher the percentage of lead the higher the price paid per unit. When the ore contained less than a certain percentage of lead, which varied with the quality of the ore, mine owners received no remuneration for lead contained in ore, however rich it may be in silver. Unless a time contract had been entered into, price varied from month to month, governed by market quotations of silver and lead at New York as well as the prices of coke, charcoal, and fluxes. When present in excess of one-tenth of an ounce to the ton, gold was paid for at the rate of $18 per ounce. Transportation of ore from mines to sampling or smelting works was paid for by the smelters at the rate of $1 to 1.85 per ton.[32] Sampling was not the only business linked to base metal smelting. Also important were suppliers of furnace components, iron fabrication, and firebrick.

In the early days of smelting, Western mining companies were handicapped by not having a source of firebrick close at hand. What refractory brick was available had to be shipped from Baltimore, Maryland, or England. Utah, however, was fortunate in that deposits of silica sand and koalin suitable for the manufacture of refractory brick were discovered about the time smelting got underway in the territory. Also present in Utah were knowledgeable individuals, mostly Mormon converts from England, who knew how to work those materials into brick. Thus was launched a firebrick industry that would grow hand-in-hand with smelting. Two Salt Lake City companies rose quickly in firebrick making: the Utah Fire-Brick Manufacturing Company, and the Pioneer Fire Brick Company. Both were formed about the same time, in 1873, coinciding with the building of smelting plants in the Jordan Valley.

The Pioneer Fire Brick Company, of Morris and Evans, fabricated firebrick from a clay obtained in Bingham Canyon. The company manufactured all kinds and sizes of brick: furnace, wedge, circle, square, bull heads, arch bricks and burrs; some of the latter being of very large size: three feet long and one foot wide. They also produced oven bottoms. In 1873 the company employed 65 men, who produced 50,000 brick, that sold for $100 per thousand. Although not Utah's largest firebrick company, Morris and Evans erected some of the largest smelters — the Germania Refining Works, the Saturn Smelter at Sandy, Carson & Buzzo's works at West Jordan, the furnaces of the Last Chance, Flagstaff, Mountain Chief, Sheridan Hill, and the Utah Queen. They also contracted to build charcoal kilns, of which more will be said.[33]

The Utah Fire-Brick Manufacturing Company, situated on the line of the Utah Southern Railroad near the sampling works of Johnson & Company, was larger than the Pioneer Company. The company was organized in February 1873 with W. W. Chisholm as president, C. W. Bennett secretary and treasurer, and W. T. Matthews, superintendent. Simon Bamberger, was superintendent in 1874. Capitalization was $30,000.

Company works covering an acre, embraced a large drying floor (110 by 35 feet), under which flues were built for drying bricks in winter time; an engine house containing a 15-horsepower stationary engine drove the machinery which consisted of two large steam vats, of a capacity of 30 tons of clay each; a pug mill; a kiln capable of burning 37,000 bricks at one time.

Crushing machinery and rollers, built by the Salt Lake Iron Company, pulverized the clay, which was subjected to an iron brick press made by Tuckfield & Sons of Salt Lake City. It was claimed that this piece of machinery was superior to imported presses. The mold in imported presses were lined with steel, while the mold of the homemade machine was simply hardened cast iron which cleared itself better.[34]

The company's clay came from a number of sources. A white fire clay of "a very peculiar and rare quality," obtained at Lehi, was deemed the best material in the world for gas retorts, a branch of business the company entered in 1874. This clay was free of iron or other "deleterious admixture." Dug out in solid, hard, dry pieces, resembling petrified wood, 45 tons of the material were consumed every week in production of 20,000 brick. Other clays were used for furnace brick.

A black clay, which came from a bed overlying a coal deposit on the Union Pacific Railroad, was used to make "K" brick. An inferior quality of clay, obtained at Lehi, went into brick used in boiler arches. A white silicate found in Bingham Canyon produced "B" brick. These bricks were moulded in all shapes according to the purposes for which they were used. There were square brick, arch, circle, key and bull head. Large, square slabs of various sizes, called "tiles," were manufactured for furnaces.

Besides making shipments to Nevada and Wyoming, the company supplied brick to most Utah furnaces. Fifteen men were employed in spring 1874, producing 75,000 brick a month. The price of bricks ranged from $80 to $100 per thousand; while brick sent to Utah from Denver, St. Louis and Chicago cost from $130 to $150. The company's reasonable prices and high quality product eliminated the importation of firebrick and other refractory materials used in furnace construction. In spring of 1880 Morris & Evans bought the Utah Fire-Brick Manufacturing Company.[35]

Smelter construction also required iron components such as boilers, engines, pumps and pipes, water-jackets, tuyeres, and crucibles, as well as fabrication and repair facilities. San Francisco, Denver, and St. Louis were a long ways away when smelting, mining and railroad companies needed fast repair services. The linking of the transcontinental railroads, and inauguration of a narrow gauge railroad system turned Salt Lake City and Ogden into centers for foundries and machine shops.

The Salt Lake Foundry and Machine Shop, also known as the Salt Lake Iron Company, situated on the line of the Utah Southern Railroad, one block from its depot, was the most complete machine shop in Utah Territory. Leased by Edmund Wilkes, an owner of the Miller Mining Company, the foundry was housed in a two-story building. Its equipment consisted of three lathes (the largest being 24 feet), upright drills, a bolting and milling machine, a small planer, and a turning lathe for pattern makers who occupied the upper story of the company's building. Machinery was driven by a 15-horsepower engine. Employing 33 men, the company turned out all kinds of iron and brass castings. Mining, milling, and smelting machinery

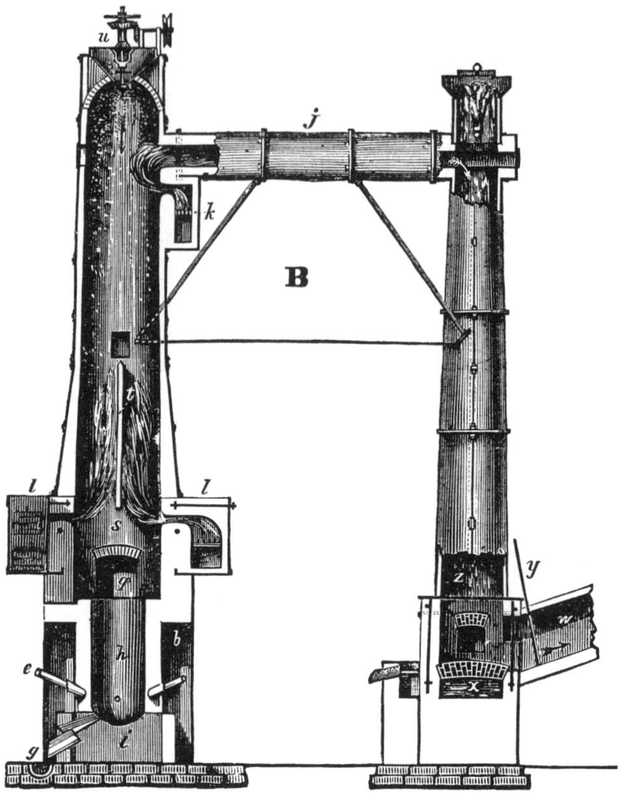

G. M. Gerrish and Philip Hinkle who had ties to George Hearst, maintained offices in Salt Lake City where they designed and built smelting furnaces such as those pictured above and at right. Their roasting and smelting furnaces, widely used in Utah, were among the first to employ automatic loading, ore being carried by a screw-conveyor *p*, from the battery to the furnace, and by an elevator, *o*, to the top of the furnace shaft. (From *Mining and Scinetific Press*, April 29, 1871)

was manufactured on the premises. The company erected quartz mills, performed narrow gauge railroad repair, fabricated hand-push freight and mining cars. It also produced ornamental house work and building fronts. In March 1874 the company erected a cupola with a capacity of three tons an hour, which enabled it to mould the heaviest class of iron castings used in the territory. It produced wrought iron and cast iron water-jackets; some of which went into the Flagstaff, Sheridan Hill, and Waterman smelters.[36] Any type of machinery was manufactured by the company, which consumed about 12 tons of pig iron a week

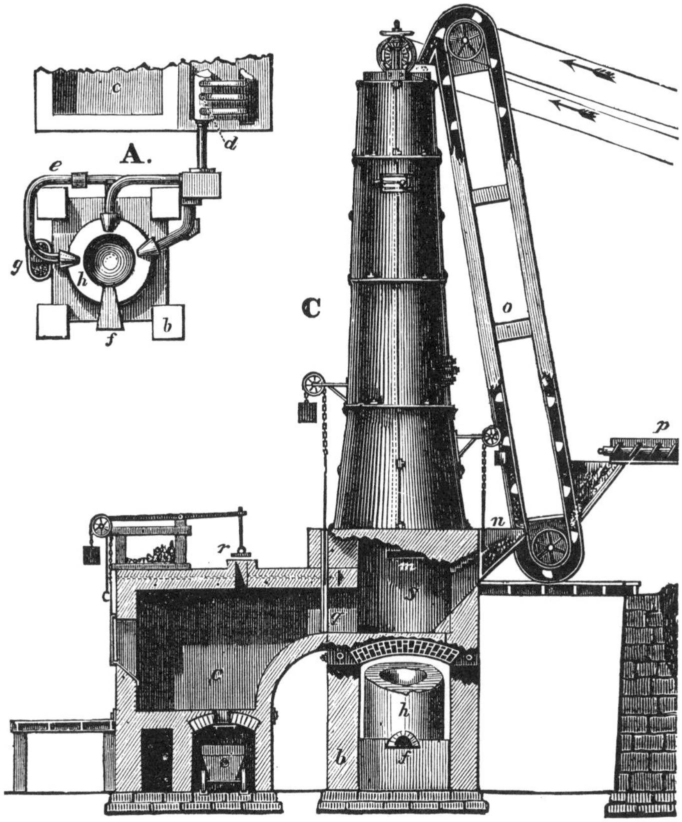

Not as large as the Salt Lake Foundry, Nathan Davis & Sons' Iron Works produced high quality components for Utah's mines and smelters. Located in Salt Lake City on First West Street, between North Temple and First North Street, the company produced the large silver refining kettles for the Germania Refining Works, and other castings such as water-jackets, slag pots, and bullion moulds. The water-jackets made by the firm were "said to be superior in construction to any that have heretofore been made in Utah. The eastern water-jackets not being considered so well adapted to the business as those manufactured here." Company machinery was driven by a Griffith & Wedge 12-horsepower engine.[37]

The Machine shop of W. J. Silver, in the Nineteenth Ward of Salt Lake City, turned out steam plunger pumps, one of which was used in the Emma Mine. And J. W. Tuckfield & Son, on State Road, created all types of brass castings.[38] The Salt Lake City Boiler Works, owned by John Sloan & Company, situated on First West Street between South Temple and First South Street, manufactured boilers,

John Williams, Jr., improved blast furnaces of the Chicago Company at Stockton with the addition of flues or dust chambers, C, C, to collect lead particles emitted from the furnace. The chambers were constructed of heavy sheet iron lined with brick, and were 12 feet long by six feet wide. (*Utah Mining Gazette*, September 27, 1873)

wrought iron furnaces for smelting silver, copper or iron, iron tanks, doors, shutters and wrought iron water-jackets. One of the largest furnaces ever made in Utah was standing on the premises in March 1874. Rollers used by the company for bending boiler plate weighed about 3,000 pounds, and were made by Nathan Davis & Son. Sloan & Company imported a punch weighing 1,900 pounds from New York, as well as shears weighing 1,100 pounds. The company employed eight men, either on the premises or as repairmen in various mining camps.[39] Now back to other smelting locales.

* * *

Stockton in Rush Valley, 40 miles west of Salt Lake City, was a natural smelting center. Like Sandy Station it was situated in the

FIG. 1.

FIG. 2.

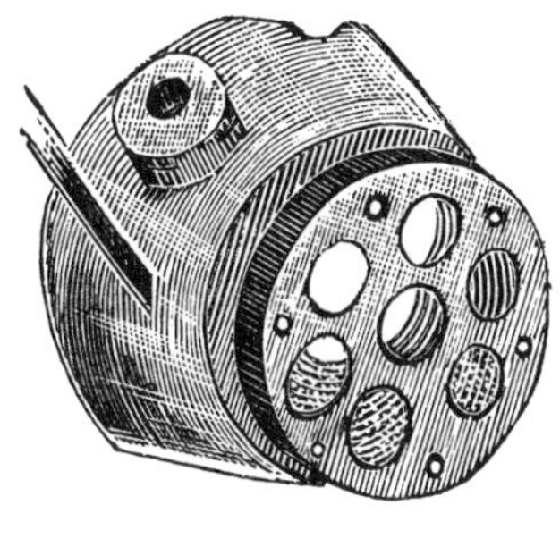

Noxious fumes were the enemy of smeltermen everywhere. In 1879 this respirator was designed and marketed for workmen engaged in lead smelting, chemical works, crushing mills, bullion refineries, and quicksilver mines. Air was inhaled through wet sponge or cotton, and expired through a valve at the side. How well it worked is anyone's guess. It was probably better than nothing. The respirator, sold by Seth Marshall, Jr., of San Francisco, sold for $3. (*Mining and Scientific Press,* April 19, 1879)

middle of a region that produced a variety of smelting ores. Most mines were in the mountains east of town, which allowed for easy transportation of ore. The silver-lead ores of Dry and East Canyons occurred in a calcareous gangue which fluxed well with silicious ores of Camp Floyd. Some Dry Canyon mines also produced argentiferous oxides of iron, which yielded silver without addition of iron ore. Lime for flux, which cost $9 per ton elsewhere in the territory, was found in large quantities near Stockton and could be laid down for 75 cents per ton, and iron ore for fluxing, as good as Wyoming hematite, was delivered at Rush Valley for $14 per ton. That commodity ran $16 to $20 per ton in Salt Lake Valley. Fuel was also plentiful and cheap, a third less than in the Jordan Valley. [40]

By 1872 Stockton had become a bustling ore reduction center. Two years later the burg's population was about 500 inhabitants, most of whom were employed in nearby reduction works. Between the mouth of Dry Canyon and Ophir City were four mills, four arrastras, and at least five furnaces. Near East Canyon, Schofield, Abbey, Drake & Company erected furnaces. Known as the East

The Chicago Compar y's furnaces at Rush Lake. Note flues above roof of smelter. (*Utah Minir ; Gazette*, June 20, 1874)

Cañon Smelting Works, they began their first run on July 10, 1871, and reduced ten tons of ore per day. Nearby was the steam powered Pioneer furnace, consisting of two stacks, having a daily capacity of three tons of bullion. Similar to the Pioneer furnace was the two stack smelter of the Ophir Mining and Smelting Company, located near Ophir in East Canyon. The first of its two furnaces was completed in July 4, 1871, under the superintendency by G. M. Gerrish. A second furnace was blown in at the end of the year.

Both furnaces were similar in design: built of eight inch thick firebrick, 14 feet from the tap to feed hole; 30 inches in diameter at the top, 40 in the middle, and 30 at the bottom. Each furnace had three tuyeres. A No. 7 Sturtevant blower driven by a 16-horsepower engine furnished the blast. The capacity of each furnace was two tons of bullion a day.[41]

One mile from Stockton, near the end of Rush Lake, Isaac S. Waterman and Henry Simons of Philadelphia erected the Waterman Smelter in May 1871 to treat ore from the Rush Valley District. The works consisted of a large round roasting furnace, 12 feet wide at the top and six feet at the bottom. Ore charges were roasted from three to six days to drive out the sulphur, and then transferred to a blast furnace, 30 to 40 inches in diameter, made of stone found nearby. In September 1871 eight tons of ore was smelted daily, the bullion sent to Omaha and Chicago for refining. Another blast furnace was built in September 1871. The Waterman Smelter was idle for sometime in 1872, but was started up in December of that year, under supervision of Henry Prosser, an experienced smelter.[42]

The fourth furnace, the H. S. Jacobs' Smelting Company, was situated a short distance east of Stockton. In June 1874 the works consisted of three vertical blast furnaces, using a 16-horsepower steam stationary engine (of Starkey's patent) as motive power. Its machinery was shipped across the Great Salt Lake by steamer from Corinne in spring of 1872, and erected under the guidance of John Williams, the father of John Williams, Jr., of White Pine and Eureka fame.

The furnaces of the H. S. Jacobs' Smelting Company were enclosed in a substantial frame building, 86 x 30 feet. Ore used at the works was obtained from the Fourth of July and Kearsarge mines in Dry Canyon. When running, the Jacobs' smelter produced about five tons of bullion a day, assaying 250 to 300 ounces silver and one and a half ounces gold per ton. No reduction works in the West, either furnaces or mills, worked ores up to a higher percentage of precious metal than did the Jacobs' Smelting Company. The bullion was shipped by lake steamer to Corinne and sold to Eastern buyers.[43]

The dominance of these two companies was challenged in spring of 1873 when the Chicago Company moved in and began construction of a new smelter. Capitalized at £150,000, divided into 15,000 shares, this English company was the brainchild of William S. Godbe, an excommunicated Mormon turned miner, who was also the company's general manager. Officers were: John H. Latey, superintendent; E. J. Dowlen, auditor; A. W. Atkin, chief clerk; John J. Williams, Jr., was furnace foreman; O. E. Avery, assayer, and L. D. Davis, foreman. Having been at White Pine and Eureka, Nevada, both Latey and Williams were knowledgeable smelters.

Under supervision of John H. Latey and John J. Williams construction of the furnace began in May 1873 within 100 feet of the east shore of Rush Lake. Completed in August, the works consisted of two vertical blast furnaces driven by a 40-horsepower horizontal steam engine built by the Utica Manufacturing Company. Steam was generated in a tubular boiler, four feet in diameter and 12 feet long. Two No. 5 Roots blowers produced the blast. Adjustable blow-off valves in the air mains facilitated the working of one or two furnaces without altering speed of the engine. The furnaces were enclosed in a substantial 75 by 48 foot frame structure with fireproof roofing. Engine and boiler house was 30 by 30 feet. The water supply for the works came from a well sunk within the furnace building.

Behind the furnaces were fuel sheds where 20,000 bushels of charcoal and 40 tons of coke were maintained at all times. Obtained under contract from adjacent mountains, the charcoal was produced chiefly from nut pine, and delivered at the works at 21 to 22½ cents per bushel. Coke was obtained from Pennsylvania for $36 to $42 per ton.

Under supervision of "Uncle John" Williams the Chicago Company's furnace ran successfully from August 1873 through 1875, being closed down only for holidays. Processing ore from the Chicago Mine and other properties in Dry Canyon, Columbus, Camp Floyd, Bingham, Little and Big Cottonwood canyons, Parley's Park, and Tintic, the furnace produced during this period 1,136 bars, or 90,880 pounds of bullion, which was sold to the Omaha Smelting & Refining Company.[44]

Interestingly, this was the first furnace in Utah to be equipped with a fume chamber, which not only captured 16,968 pounds of lead-rich dust every seven days, but preserved the health of its employees. Which explains why the furnace ran continually for seven months without any cases of "lead cholic during the whole time." The good health of the men was attributed to "excellent arrangement for ventilating the works . . . and the thoroughly practical nature of the fume condensing chamber attached, which utilizes the lead ejected from the stack, instead of allowing it to go to waste on the roof and ground, poisoning all it comes in contact with." A fume chamber of the same type was later adopted by the Winnamuck works in Bingham Canyon.

Running for a number of years, the Chicago Company was the most successful smelter in the district. In 1876 it doubled its size by purchasing the Longmaid smelter, also located at Rush Lake. By then, however, most of the smelters at Stockton had fallen by the wayside, the *Salt Lake City Tribune* stating on July 12, 1876, that the district had only three smelting companies in operation.[45]

"Uncle John" Williams, superintendent of the Chicago Company smelter and the most knowledgeable metallurgist at Stockton, was not present when the Chicago Company purchased the Longmaid smelter. In 1874 he and Latey moved south to Diamond City in the Tintic District to erect and manage for a short time the Germania Smelting Company's furnace. Not to be mistaken for the Germania Refinery at Murray, this smelter consisting of one vertical blast furnace, was designed to reduce copper carbonates from the Mammoth Mine.

Although only a single furnace, the Germania smelter was of large capacity, reducing 24 tons of ore daily. By all accounts the smelter was well managed, John Williams bossing a crew of 14 men, who kept 14 mule teams employed transporting ore and matte. It was a busy operation, eating up 600 bushels of charcoal every 24 hours. That commodity was furnished by Charles Lamendorf, a Tintic general merchandise and mining supply dealer, who had branched into the fuel business. To furnish district smelters with a superior grade of charcoal, he erected a number of charcoal kilns near Homansville. When the Utah Smelting and Mining Company[46] and other smelters in the district shut down in 1874-75 due to lack of smelting ore, Lamendorf's charcoal went exclusively to Latey and Williams.[47]

John Latey and John Williams did not stay long in the Tintic District. Discovery of large deposits of argentiferous lead drew them to Beaver County, 242 hundred miles to the south. But first some background.

In the foothills of the eastern slope of the San Francisco mountains, 50 miles west of Beaver, Utah, and 90 miles from Pioche, Nevada, Sam Hawkes and James Ryan, the latter a Beaver merchant, located a ledge of silver-bearing galena in 1871. They filed the Grampian claim which drew other prospectors. The following year more claims were staked, some rich in free-milling gold and silver ore which was shipped to Pioche and Salt Lake City for processing. A camp called Grampian City sprouted on the west side of the range overlooking Wah-Wah Valley, and the San Francisco Mining District was off to a fair start.

On September 24, 1875, while working their claim, Hawkes and Ryan discovered more silver-bearing ore 750 feet to the northeast and 200 feet higher on the hillside. Because of a high concentration of silver chloride or cerargyrite, popular called horn silver, they

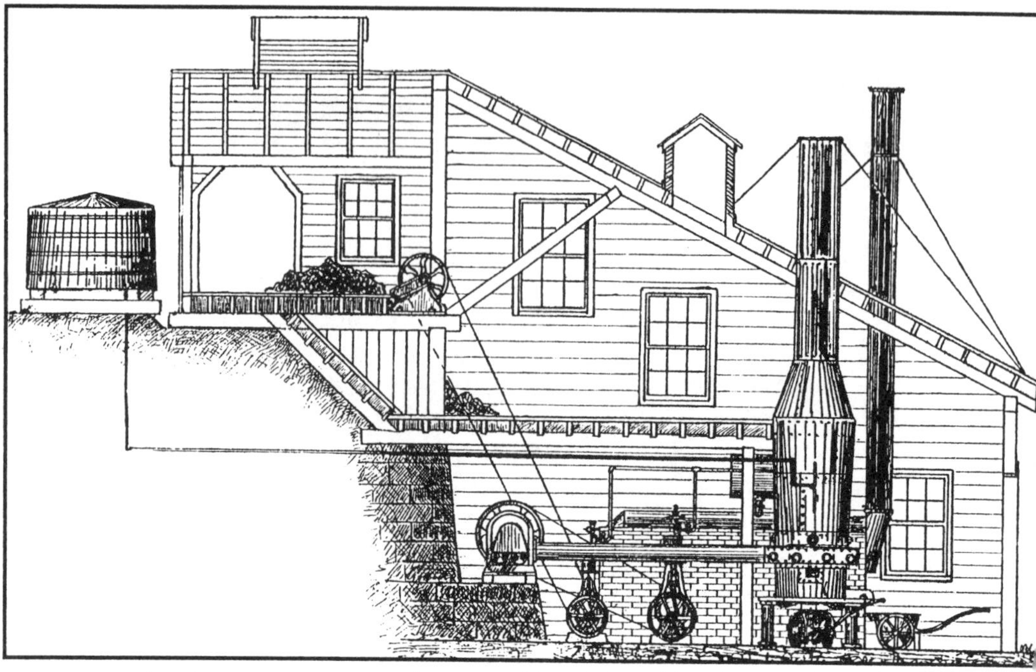

Smelters were constructed on several levels. A rock breaker was located on the first level, upper left, through which ore was run. The pulverized ore was received on the next level, center, where it was mixed with flux and fuel, either charcoal or coke, and then fed to the furnace on the lower level, right. Blast was supplied to the furnace by blower in center of illustration. A tank at upper left held water for cooling the furnace. A boiler and steam engine, in background, powered the entire plant.

named that claim the Horn Silver. Hawkes and Ryan sunk a shaft 30 feet into the ore, and a few months after the discovery, in February 1876, sold for $25,000 their right, title and interest in the Horn Silver claim to Allen G. Campbell, Matthew Cullen, Dennis Ryan, and A. Byram.

Campbell, Cullen and associates proved the deposit to a depth of over 280 feet. It was an immense orebody, far larger than the Grampian. On the surface the vein was traceable for two miles, with a general course of north 10 degrees west. The ore was both milling and smelting grade. The latter, consisting of sulphate, oxide and carbonate of lead, was soft and earthy; high in silver content. It predominated over the milling ore. A shaft worked by a two-horse whim followed the orebody downward for three levels; on the third level the ore was 300 feet wide. Heralded as Utah's greatest bonanza, the Horn Silver lode put the San Francisco District on the map.

Because this ore was heavily charged with lead (30 percent) and contained from 30 to 100 ounces of silver to the ton, Campbell, Cullen and associates needed a smelter. With the aid of W. S. Godbe, they purchased in spring of 1876 the Holcomb smelter at Wah-Wah, a draft furnace built in 1874 to work ore from the Mountain Queen claim, which had failed for want of funds. They moved the smelter to the San Francisco Mining District, and built five charcoal kilns closeby to supply fuel.[48]

Campbell & Company's furnace, also called the Frisco Smelter, was constructed of local rock and had a flue extending up the hillside several hundred feet, which was hoped would create sufficient draft to dispense with an engine and blower. That was not to be the case. When fired up in early July 1876, a few tons of bullion was made, but the draft could not be regulated and the furnace froze up. A No. 5 Baker blower driven by a 40-horsepower horizontal engine was installed, and by September the single furnace was reported "running all right," turning out from 80 to 90 bars of bullion every eight hours.[49]

Another town, appropriately christianed Frisco, took root near the smelter about two miles from the Horn Silver Mine. The burg ultimately attained a maximum population of about 600. The town had one great drawback, a scarcity of water. Although the smelter drew its water from a well drilled for that purpose, all water for domestic use was hauled in from Black Rock Station, 40 miles distant, and retailed at seven cents per gallon. Perhaps that is why Frisco had 30 "gin mills," in contrast to one "gospel mill." It did develop a number of substantial business houses, including the Southern Hotel, a large and neatly furnished house built of fire-proof sandstone.[50]

While the Frisco Smelter was being erected, John Latey and John Williams made their appearance in Beaver County. In fall 1876 at Riverside Station on Beaver Creek, a few miles from Frisco, they erected a quartz mill and custom smelter to reduce "copper and all rebellious ores." A small town, later called Milford (a combination of the words mill and ford), was laid out at the river crossing near the mill and smelter.[51]

By mid-September 1876 the mill at Riverside had run out its fifth silver brick. Ninety days later the furnace of Latey and Williams was in operation. Because it reduced ore from the Horn Silver Mine, this furnace was frequently labeled the Horn Silver smelter, although it processed ore from other mines in the district. Fuel for the smelter came from five charcoal kilns near Copper Canyon, owned by the Liddel Brothers, Gampian City merchants. To keep pace with production, they added two more kilns. All this activity caused the *Tribune* to speculate on September 12, 1876, that "big and busy times are anticipated in Beaver county this fall." The newspaper was right.

During the 1876 season the Frisco smelter ran day and night, sending between 40 to 50 wagon loads of bullion to the railroad terminus at York. A second furnace was planned, but before it could be built the original Holcomb smelter burned down in August 1877. It was replaced with a 30 ton furnace, and by January 1878 the smelter was again in operation.[52]

To manage the Frisco smelter, Campbell & Company brought in Franz Fohr. Although quiet and unassuming, Fohr knew the smelting business. According to the *Engineering and Mining Journal,* he was "practically the last of a galaxy of metallurgists who created the art of silver-lead smelting in this country between 1870 and 1885." Fohr's ancestry, education, and early career, however, is somewhat of a mystery. When he came to America is not known. What is known is that Franz Fohr was born on September 7, 1838, in Mannheim, Germany. By 1870 he was an experienced metallurgist, for he was superintendent of Edward Balbach's Newark Smelting and Refining Works from July 1870 to January 1, 1872. Shortly thereafter he left Newark for San Francisco, where he was commissioned by the Thomas H. Selby & Company to compile information on the manufacture of white lead. With termination of that commission on January 1, 1875, Franz went to Silver Islet in the Lake Superior copper region. It is not known what he did there. In early 1876 he was associated with the Boston Silver Company operating at Saint John's, Summit County, Colorado, smelting the company's lead ore. That job ended in summer of 1878 when Fohr took charge (on August 1, 1878) of the Frisco smelter, an operation he conducted until August 1, 1879.[53]

In the meantime the Horn Silver Mine increased production, and by September 1876 it was able to furnish enough ore for Williams and Latey to run two stacks, and the next two years were good for these smeltermen. Their prosperity, however, would be impacted by what Utah newspapers called the onset of "big times."

The "big times" began in February 1879 when C. J. Francklyn of the Cunard Steamship Line, bankers Jay Cooke and Frank G. Brown, and other investors of New York and eastern cities, purchased a half interest in the Horn Silver Mine from Campbell, Cullen, and associates. Not included in the deal was the Frisco Smelting Company, a stone quarry, wood ranches, five charcoal kilns, barns, wagons, and animals. The Horn Silver Mine was capitalized at $10,000,000, divided into 400,000 shares. Immediately thereafter Jay Cooke persuaded the Utah Southern Railroad, which had taken seven or eight years to creep 75 miles south of Salt Lake City, to build to Frisco.[54]

In February 1879 the Frisco Smelting Company, owned by Campbell, Cullen, Godbe, and others, consisted of two furnaces, situated above the town. Regardless of the difficulty in obtaining water, these furnaces ran successfully and paid well. In late September 1879 the smelter owners consolidated their properties with a number of mines which they owned: the Carbonate, Cave and its group in the Bradshaw District, the Rattler Mine, and an iron flux mine in the Rocky District, to form the Frisco Smelting & Mining Company, capitalized at $2,000,000, divided into 80,000 shares. Messrs. Bigelow and Rasin, Philadelphia financiers, became large shareholders in the company.[55]

Frisco and Milford boomed a year later when the Utah Southern line was extended to the Horn Silver Mine. Price of water dropped to four cents a gallon at Frisco, and property values skyrocketed at Milford. Owing to lack of water and growing scarcity of fuel and flux smelting at Frisco was a "hard formation," as one reporter put it. The Horn Silver Company settled this dilemma by resolving to move its smelting works north. In 1880 it purchased 40 acres on Big Cottonwood Creek, six miles south of Salt Lake City, close to the line of the Utah Southern Railroad, a mile and a half south of the Morgan Smelter. The place was christened Francklyn in honor of the company's president.

The Horn Silver Company hired August Raht, a German metallurgist, to build a smelter on the Salt Lake Valley site. Known as the "Prince of lead smelters." August Raht was born February 25, 1843, at Dillenburg, Duchy of Nassau, Germany, where his father was president of the Court of Appeals. After passing through the gymnasium, the equivalent of prep school, he studied mining and metallurgy, graduating from the Polytechnic Institute at Hesse-Cassel in 1863, and continued his studies at Freiberg. He came to the United States about 1867, taking a position with the Ducktown Smelting and Reduction Company, at Ducktown, Tennessee. Except for a two year sojourn to Italy and Africa, Raht worked for this smelting company until he went West in the early 1880s to take a position with the Mingo smelter in Utah's Salt Lake Valley.

Raht applied his skill and expertise, and by end of 1880 an enormous smelter took shape. Blown-in on June 11, 1881, the Francklyn smelter was the largest works of its kind in the West.[56] Designed by Joseph M. Locke, and manufactured by the Lane & Bodley Company of Cincinnati, Ohio, the plant consisted of five blast furnaces, each with a daily capacity of 60 tons. Costing $160,000, this was the finest refinery of its time, "complete with every modern appliance . . ., [and] lighted with the electric light." It ran around the clock, employing more than 70 men. Average monthly output was 875 tons lead, 140,000 ounces of silver. The bullion was not refined but shipped in bars to Chicago where the company had its own refinery near 39th and Clark streets on the city's south side.[57] There, the silver was separated from the lead. The Horn Silver Company also had "an experimental smelter," where new "theories" were tested, on the west side of Chicago near 42nd Street.[58]

For a few years the Horn Silver Mine produced fabulous amounts of argentiferous lead ore, giving employment to about 150 men. The mine, however, was unlike any in the west, being virtually dry. The yellowish ore, about 30 percent lead, came out dry and semi-pulverized, making the Horn Silver Mine an extremely unhealthy place to work. According to the *Mining and Scientific Press,* workmen inhaled or absorbed the lead powder to such an extent that "many miners . . . left the locality 'leaded' so badly that they were partially paralyzed." For that reason Horn Silver workmen were "more than ordinarily temporary, the force of undergound miners . . . constantly fluctuating." Nevertheless, the Horn Silver Mine and smelter combined to make it the "most productive piece of mining property in the United States."[59]

While the Francklyn Smelter was good for the Horn Silver Mine, it had a detrimental affect on the town of Frisco. With shipping of ore to Salt Lake Valley, it was no longer necessary to transport fuel, flux, and other materials to the district. The old Williams & Latey two stack smelter at Milford was blown-out on June 27, 1882, and dismantled.[60] The Frisco Company likewise ceased smelting, preferring to send its ore to the Francklyn Smelter. The consequence was that the town of Frisco suffered a loss of 300 to 400 people.

The underground work sustained the town for another year, the Horn Silver Mine sending 140 tons of ore daily to the Francklyn

Joseph Locke, Salt Lake City manager of the Lane & Bodley Company, Cincinnati, Ohio, furnace builders, designed and built this lead furnace for the Francklyn Smelter of the Horn Silver Mining Company. Measuring 40 inches by nine feet at the tuyeres, five of these gigantic Raschette furnaces produced over $2,000,000 worth of bullion for the Horn Silver Mining Company during the first six months of 1882. (*Mining and Scientific Press*, January 27, 1883)

Smelter.[61] And then it all came to an end on Wednesday night, February 11, 1885, when the great Horn Silver Mine caved in. According to the *Tribune* of February 14:

> The men came out of the mine as usual for supper, and returned below at 12 o'clock. Before they got distributed in the mine, they became alarmed and came to the surface. They then went back again to investigate the situation, and the movement of the mine below the 300 foot level frightened them out and they were hoisted to the surface as fast as possible. When the last cage, on which six men were coming up, had reached a point about 200 feet below the surface, the shaft closed in so as to catch the cage and hold it, and the men had to be pulled up, one at a time, by a rope let down to them. They all got out rapidly and without injury to any.

The collapse was not unexpected. For sometime before the cave-in there had been movement of strata within the mine, which snapped mine timbers as if they were match sticks. Nevertheless, the disaster had lasting consequences. By February 24, 1885, over 100 men had been discharged from the Horn Silver; 20 to 30 men being kept on to assess the situation in the mine's interior. For months no news was published about the mine. When another shaft was sunk and some production resumed, the *Tribune* proclaimed the Horn Silver "still a great mine." That was premature, for on Sunday night, April 5, 1885, "with a mighty crash," the old workings caved again. Although no one was injured, the collapse sealed an air passage in the new workings "so that . . . a match could not be lighted on the seventh [level] in the new shaft." The *Southern Utah Times* was quick to note that "if the previous cave did not put an effectual kibosh upon the old shaft and works this one has."

The paper was right, it was "the straw that . . . everlastingly broke the business back of Frisco." It also had repercussions in Salt Lake Valley. With no production coming from the mine, the Francklyn Smelter was blown-out at end of March and 14 carloads of bullion, valued at $600,000, moved to market.[62]

Collapse of the Horn Silver Mine did not put the company out of business. It sank a new shaft, repaired its underground workings, and limped along. A small amount of ore was produced in 1888. A year later 100 men were at work and about 100 tons of ore per month was shipped to Salt Lake Valley smelters. In 1889 the mine paid a dividend of $50,000.

About 160 men were employed in 1893, and in early 1894 the mine built a concentrator close to the shaft. Then disaster struck again. On April 4, 1894, a fire destroyed the hoist, shaft, and concentrator, drastically curtailing production. Resilient as ever, the old producer repaired its workings, and by January 1, 1896, the Horn Silver Mine was in operation again. During the next few years its shaft was deepened to 1,600 feet and about 1,000 feet of drifting done. In 1897 the mine was declared in a "prosperous condition," a statement not quite true. Although the Horn Silver Mine had paid $5,300,000 in dividends by 1900, argentiferous lead ore was growing lean. A shoot of copper, struck years before, was investigated, but there was not enough good ore remaining to keep the mine open and it closed in 1900.[63]

* * *

Utah's smelting industry started to decline in 1877, due mainly to a drop in the price of silver, which forced the closure of many silver mines. In January of that year there were ten smelters in Utah: two in Tooele County, two in Beaver County, and six in Salt Lake County. Ninety days later the Saturn and Galena furnaces, "old and dilapidated," suspended operations. Competition from eastern and midwest smelters idled two more smelters. In 1884 Sandy Station was reported "dull." Collapse of the Horn Silver Mine and cessation of its Francklyn Smelter marked the end of the heyday of lead reduction in Salt Lake Valley.

Smelters that survived were those not tied to mining companies. Known as custom smelters, they drew ore from a variety of producers, not only in Utah but from as far away as Nevada, Idaho, and Montana. There were only three. The Germania Refinery ran throughout this period, and enlarged its operations, adding a white lead department in 1882. The Morgan Smelter — renamed the Hanauer in 1881 — was going in 1884, running two out of its five furnaces.[64] The Mingo Furnace Company, consisting of four stacks and two large reverberatory furnaces, ran continuously except for 90 days in spring 1888 when it was rebuilt following its destruction by fire in February. It was enlarged in 1892 with the addition of two 36 x 120-inch blast furnaces. The continuation of these three smelters, of course, kept the Valley's three sampling firms in operation: Conklin Sampler, Sandy Sampling Mill, and the Pioneer Sampler.[65]

For 13 years the Germania, Mingo, and Hanauer were the only custom lead smelters operating in Salt Lake Valley. Then in 1897 a fourth smelter, the Pennsylvania Lead Company or Pennsylvania Smelting Company, entered the act, building a custom smelter at Sandy in 1897. This company, which had a lead refinery at Allegheny City in western Pennsylvania, had been around for a long time. Established by Meyer Guggenheim and sons, the company in summer of 1882 built a two-furnace smelter near Ketchum, Idaho, to process silver-bearing lead produced in the Wood River District. Four more furnaces were added, and the Pennsylvania Company grew into a major concern building another facility at Pueblo, Colorado, in 1889. It was only natural they would eye the Salt Lake City region, lying as it did in the midst of a lead-producing region, as a location for yet another custom plant.[66]

A major change was underway to Utah's smelting industry by the late 1890s. Mining of lead was giving way to copper mining, and that would cause the demise of the old historic lead smelters. That story began in Bingham Canyon at the Highland Boy Mine.

In 1896 Salt Lake City banker Thomas Weir and mining promoter Samuel Newhouse bonded considerable mineral-bearing property near the head of Carr Fork, which included the Henry M and Omaha claims, as well as the Highland Boy Mine, the latter located in 1873 by J. W. Campbell. Without capital claims cannot be turned into paying mines, and Samuel Newhouse set out to attract money, a job he was proficient at.

Born about 1854 in New York City of Russian-Jewish parents who came to the United States in 1829, Samuel Newhouse studied law, was admitted to the Pennsylvania bar and practiced law in Philadelphia and Scranton. Scent of profit lured him to Leadville, Colorado, where he engaged in freighting until 1886, then he turned to mining, acquiring the Wheel of Fortune Mine and other claims. A shrewd speculator and miner, Newhouse made money and by the 1890s he could count the Prince of Wales and other British nobility as friends.

Because of his ties to English royalty, Newhouse easily picked the London mining market. He incorporated the Highland Boy Gold Mining Company with himself as president and Weir as general manager, and secured £300,000 running capital from the sale of stock. Systematic exploration proved the value of the mine lay not in its lead reserves but in its oxidized zones, which contained an estimated 50,000 tons of ore, carrying from $30 to $40 per ton in gold. A stamp mill and cyanide plant were erected and production got under way. The ore, however, carried a high percentage of copper, which caused abnormal consumption of cyanide; in fact, copper was everywhere in the mine.

A large shoot of copper sulphide had been exposed in a winze off the fourth level, a find quickly sealed off for fear news of the discovery would adversely affect stock sales. Unable to efficiently process its gold ore, the company was in serious financial trouble by fall of 1896 with mounting bills and a delinquent payroll. In desperation, Samuel Newhouse went to Denver to raise money to pay production costs; and while conferring with bankers he received a telegram from Utah stating that a chimney of copper sulphide 350 feet long and 200 feet thick, assaying 15 percent, had been encountered in a lower adit.

There was no hiding that vein of copper. It was magnificent, "greater than imagination can depict." One old miner reported to the *Tribune* of September 12, 1897, that "there were millions in sight." It mattered not whether he meant tons or dollars, it all boiled down to the same thing. Bingham Canyon — and Utah — was on the eve of a copper boom.

The Electric Age was dawning and a communication network, linked by copper strands, was spanning the continent. Railroads and general construction also cried for the red metal. Copper rose to 12 cents a pound and there was every indication that the price would go higher. Newhouse intuitively knew it would be more profitable to mine copper than gold, and in December of 1896 the Highland Boy made its first shipment of copper ore. Reaction from the mining fraternity was immediate. Bingham was overrun with speculators looking for copper mines. Within two years the Highland Boy Mine had been absorbed by the Utah Consolidated Mining Company, Ltd., a New Jersey corporation. The cyanide mill was abandoned and an elaborate production facility installed, including a 12,700-foot-long Bleichert aerial tramway to transport ore from mine entrances to the railway terminal in lower Bingham Canyon.

In 1899 the company built the Highland Boy smelter on 56 acres on a bluff overlooking the Jordan River, southwest of Murray. That year the mine again changed hands, the controlling interest being purchased for $12,000,000 by the Standard Oil Company of New

York and Boston — William Rockefeller had become chief stockholder of the Highland Boy Mine.[67]

Developments at the Highland Boy Mine stimulated the search for copper. Properties with the least hint of copper were secured, often at inflated prices, and existing companies expanded production. The Bingham Copper and Gold Mining Company enlarged its Dalton and Lark mines, built a smelter at Jordan, and completed the Dalton and Lark Railroad connecting its mines with the Denver and Rio Grande Railroad at Dalton Junction. By 1900 three giant corporations — the American Smelting and Refining Company; the United States Smelting, Refining and Mining Company; and the International Smelting and Refining Company — were vying with one another in a race to gobble up Bingham Canyon mining properties and dominate Salt Lake Valley smelting. The American Smelting and Refining Company (ASARCO) emerged the winner.

Organized under the laws of New Jersey on April 4, 1899, American Smelting and Refining Company was the creation of Henry H. Rogers, Leonard Lewisohn, and John Moore. The declared powers and purposes of the corporation were "to acquire, deal in, sell, and otherwise dispose of ores, minerals, metals; and to manufacture, acquire, deal in, sell, or otherwise dispose of the products of ores, minerals, and metals." The real aim of this company was to "restore order, stability, and profitability in an industry that had been plagued during the 1890s by an over-capacity resulting from numerous smelters throughout the country competing with each other for a limited supply of ore."[68]

The enterprise was gigantic. The company was capitalized at $65,000,000, divided into 650,000 shares of stock, par value of $100 each. With that kind of financial backing the company, or trust as it was erroneously called,[69] acquired ownership of or managerial rights to 17 corporations and one partnership, consisting of 16 smelting plants, 18 refineries, and numerous mines and mining claims.

Properties comprising the company consisted of the Omaha and Grant, with a refinery at Omaha and smelters at Denver and Durango in Colorado; the Consolidated Kansas City, with a refinery at Argentine, Kansas, and smelters at El Paso, Leadville, and El Carmen, Mexico; the National United Colorado, with smelters at East Helena, Great Falls, and Pueblo, and a refinery at Chicago; the Aurora American, with refineries at Chicago and a smelter at Leadville; Pennsylvania Lead Company, with a refinery at Allegheny City, Pennsylvania, and a smelter at Salt Lake City; the Globe Company, with a smelter and a refinery at Denver; the Ibex and Bi-Metallic, each owning smelters, the former at Leamington, Utah, and the latter at Leadville; Pueblo, with a smelter and a refinery at Pueblo; the Germania Refinery, the Mingo and Hanauer smelters, all in Salt Lake Valley. And on April 8, 1901, ASARCO acquired the properties of M. Guggenheim and Sons.[70]

It was easy for Rogers and Lewisohn to put together a giant corporation. To make it function profitably was a different matter. Top-heavy with plants, the organization was pruned to reduce unnecessary expense. Smelters were closed at Denver, Great Falls, Leadville, and Kansas. In Salt Lake Valley the Pennsylvania Smelting Company, and the Hanauer and Mingo smelters were shut down in April 1899. And in 1902 the great Germania Refinery was closed upon completion of ASARCO's new $1 million lead-silver smelter at Murray.[71]

An era ended with closure of the Germania Refinery. For 30 years Salt Lake Valley had been the scene of metallurgical experimentation. Along the Jordan River, south of Salt Lake City European expertise and innovation had combined with Yankee ingenuity to create a great industry. Except for a few down periods, furnaces ran around the clock, lighting the night sky and clouding the day with smoke and haze. The emissions were noxious to be sure. The fumes, laden with particles of lead and arsenic killed crops, poisoned livestock, smeltermen, and local residents. Few complained, for that was progress. After all, production of bullion put Utah on a sound financial footing. Salt Lake City became the first point in the United States where a general smelting business of magnitude came to be conducted; its practice the prototype after which that of Denver, Pueblo and El Paso was fashioned.[72]

* * *

1. *Salt Lake Daily Herald*, July 24, 1875.

2. *Mining and Scientific Press*, April 24, 1880; Lynn R. Bailey, *Old Reliable. A History of Bingham Canyon, Utah.* Tucson: Westernlore Press, 1988, pp. 12-17; Lynn R. Bailey, *A Tale of the "Unkilled." The Life, Times, and Writings of Wells W. Spicer.* Tucson: Westernlore Press, 1999, pp. 18-19; B. S. Butler, G. F. Loughlin, V. C. Heikes and others. *The Ore Deposits of Utah.* United States Geological Survey, Professional Paper 111. Washington: D.C., Government Printing Office, 1920, pp. 115-16, 254, 362-63.

3. *Engineering and Mining Journal*, August 15, 1871, p. 105-6.

4. *Ibid.*, October 10, 1871; Ellsworth Daggett, "Economical Results of Smelting in Utah," *Transactions*, American Institute of Mining Engineers, Vol. II, May 1873-February 1874, p. 20.

5. The *Mining and Scientific Press,* September 2, 1871; H. Hahn, "Notes on Lead Smelting in Utah and Nevada," *Engineering and Mining Journal,* February 13, 1872, pp. 97-98; *Tribune,* May 1, 1871.

6. *Tribune,* November 22, 1872; *Mining and Scientific Press,* September 20, 1873; January 27, 1877.

7. *Mining and Scientific Press,* September 2, 1871; January 11, 1873; *Tribune,* July 14, 1871.

8. *Tribune,* November 19, 1872.

9. *Ibid.,* July 15, 1871; *Utah Mining Gazette,* November 29, 1873.

10. *Tribune,* January 2, 1875.

11. *Ibid.,* November 22, 1872.

12. The American Fork Railroad ran off and on until May 1878, when cessation of mining at the Miller Mine closed the line. Track was removed the following month. Clarence A. Reeder, Jr., *The History of Utah's Railroads 1869-1883.* New York: Arno Press, 1981, pp. 193-208.

13. *Tribune,* January 6, 1873.

14. *Ibid.,* October 16, November 24, 29, 1873.

15. *Tribune,* February 28, 1874.

16. *Ibid.,* January 6, April 21, 1873.

17. George W. Hilton, *American Narrow Gauge Railroads.* Stanford: Stanford University Press, 1990, p. 530.

18. *Mining and Scientific Press,* December 21, 1872.

19. J. C. Cameron, Jr., "Bingham Canyon. Its Mines and Reduction Works," *Utah Mining Gazette,* April 25, 1874; *Tribune,* April 29, 1873.

20. *Tribune,* January 19, 1875.

21. *Mining and Scientific Press,* January 27, 1877. Born June 20, 1833, in Raymond, Maine, Liberty E. Holden graduated from the University of Michigan in 1858. He became superintendent of schools in Tiffin, Ohio, and was admitted to the Ohio bar. He went to Cleveland in 1866 and entered journalism as owner-editor of the *Plain Dealer.* At the same time he dabbled in mining, assuming the managership of the Pittsburgh & Lake Angeline mines in 1872. Two years later he and his son, Albert, became interested in silver mines at Bingham Canyon, Utah, which engaged their attention for almost 20 years. Liberty Holden died near Mentor, Ohio, on August 26, 1913. *Engineering & Mining Journal,* September 6, 1913, p. 469.

22. *Tribune,* April 21, 29, 1873; *Utah Mining Gazette,* November 29, 1873.

23. *Tribune,* April 29, 1873; January 19, 1875.

24. *Ibid.,* April 21, 1873; *Mining and Scientific Press,* January 27, 1877; *Tribune,* January 1, 1882.

25. Mathier and Geist erected a third furnace in 1873. The smelter was shut down in 1875, while the old engine was replaced with an 80-horsepower unit, fume condensers added, along with a reverberatory furnace for roasting matte and refractory ores. New ore platforms were constructed and the rail switches modified. Two large reservoirs were built so the works would have water in any weather. All four furnaces were fired up in late 1876. *Mining and Scientific Press,* January 11, 1873; January 27, 1877; *Tribune,* January 19, 1875.

26. *Utah Mining Gazette,* May 2, 1874.

27. *Mining and Scientific Press,* January 27, 1877. For a discussion of early-day treatments of lead poisoning. done of which were effective, see Dr. William Pepper's article, "Lead Poisoning," *Scientific American,* Supplement 235, Vol. X, July 3, 1880, pp. 3745-46.

28. *Utah Mining Gazette,* November 15, 1873; February 21, 1874.

29. *Ibid.,* February 21, 1874.

30. Both Gustav Billing and Frederic Eilers went on to other enterprises. In 1879, in partnership, they built and operated for several years the Arkansas Valley smelting works at Leadville, Colorado, which ran continuously for more than 31 years. Recognizing that custom smelters must be established at railroad centers, in early 1882 Eilers organized the Colorado Smelting Company at Pueblo, Colorado. When great lead deposits in Idaho were opened, Eilers focused on the northwest in 1888. Organizing the Montana Smelting Company, he built a large plant at Great Falls, Montana. This was merged with a smeltery at East

Helena and the National Refinery at Chicago, under the name of the United Smelting and Refining Company. All of these plants were incorporated into the American Smelting and Refining Company upon its organization in 1899. Frederic Anton Eilers died at his home at Sea Cliff, Long Island, on April 21, 1917; survived by his wife, two daughters, and a son. The son, Karl, achieved a reputation worthy of his heritage, becoming vice president of the American Smelting and Refining Company. *Engineering and Mining Journal*, January 22, 1910, pp. 202-203; April 28, 1917, pp. 762-64; *Mining and Scientific Press*, April 28, 1917, p. 567-68; *Tribune*, January 1, 1875.

31. *Tribune*, August 31, 1873.

32. *Utah Mining Gazette*, September 6, 1873; May 16, 1874; *Tribune*, January 1, 1875; *Mining and Scientific Press*, September 9, 1871; January 2, 1885, p. 7.

33. *Utah Mining Gazette*, March 14, 1874.

34. *Ibid.*, September 13, 1873; *Tribune*, May 20, 1874.

35. *Utah Mining Gazette*, May 23, 1874: Emmons and Becke, *Precious Metals*, p. 438.

36. *Tribune*, July 10, 1873; *Utah Mining Gazette*, March 14, 1874.

37. *Utah Mining Gazette*, March 14, 1874.

38. *Ibid.*, March 14, 1874.

39. *Ibid.*, March 14, 1874.

40. *Mining and Scientific Press*, March 15, 1873.

41. *Utah Mining Gazette*, June 6, 1874; *Mining and Scientific Press*, September 23, 1871; January 11, 1873.

42. *Mining and Scientific Press*, September 23, 1871; December 14, 1872.

43. *Ibid.*, May 11, 1872; December 14, 1872; *Utah Mining Gazette*, July 6, 1874.

44. *Utah Mining Gazette*, September 27, 1873; *Tribune*, January 1, 1875.

45. *Utah Mining Gazette*, February 14, 1874; *Mining and Scientific Press*, January 27, 1877; *Tribune*, July 12, 1876.

46. The Utah Smelting and Mining Company works, erected in June 1871, consisted of two vertical blast furnaces of 20 ton per day capacity each. They worked ore from the Scotia Mine at West Camp. Incorporated in Utah, the company was run by John Lawrence, president, and T. S. Clarkson, treasurer. The Tintic Silver Mining and Smelting Company had similar works near Diamond City. Those works, erected in October 1871 by Hopkins, Parsons & Company, consisted of two furnaces of 20 tons per day capacity each. *Mining and Scientific Press*, January 11, 1873; *Tribune*, March 31, 1875.

47. *Tribune*, March 31, 1875.

48. *Mining and Scientific Press*, June 7, 1879.

49. Dept. of the Interior, U.S. Census Office. *Statistics and Technology of the Precious Metals*. Vol. 13., p. 469-70.

50. *Tribune*, July 1, 9, September 6, 1876; February 16, 1879; *Mining and Scientific Press*, April 9, 1881.

51. *Tribune*, September 1, 6, 1876.

52. *Tribune*, January 1, 1885.

53. Leadville being at the height of its boom in 1879, Franz Fohr moved to Colorado, becoming part owner and manager of the Malta Smelting Company, below California Gulch at Leadville. In 1881 he was superintendent of the Harrison Reduction Works of the St. Louis Smelting and Refining Company, a position he retained until retirement in the 1890s. He then moved to New York City, where he maintained an office with Faber du Faur, one of his early friends. He died at Karl Eiler's Sea Cliff residence on July 27, 1919. Fohr's modesty and his disinclination to form a wide circle of friends prevented him from attaining the fame that might easily have been his if he cared for it. *Engineering and Mining Journal*, November 22, 29, 1919, pp. 828-29.

54. *Tribune*, February 16, 1879; January 1, 1880.

55. Emmons and Becke, *Statistics and Technology of the Precious Metals*. Vol. 13., p. 469-70.

56. *Tribune*, January 1, 1885; *Mining and Scientific Press*, April 9, 1881. The *Tribune* of May 25, 1881, gave the following description of the Francklyn smelter. "Close to the track are the offices and a building enclosing the furnaces and the apparatus used in assaying. Some fifty feet to the rear of these are the smelters. There are five smelters in the works —

five superb furnaces provided with all the latest improvements, and so arranged as to do all the work possible in the best possible manner. Back of the furnaces, in a separate room, are the blasts — one for each furnace. Passing from the blast room to the left, one reaches the battery, which consists of three mammoth boilers. On one side of the works is an incline arranged for the dump, and on the the other a large level tract of ground intended for railroad tracks and apparatus to be used in transporting ore from the cars to the works."

57. *Tribune*, January 1, 1882; *Mining and Scientific Press*, January 27, 1883; November 8, 1884.

58. *Tribune*, February 13, 1881.

59. *Mining and Scientific Press*, November 8, 1884, p. 294.

60. *Tribune*, January 1, 1885.

61. *Mining and Scientific Press*, September 13, p. 166; November 1, 1884, p. 281.

62. *Tribune*, February 14, 24, June 17, 1885; *Southern Utah Times*, April 8, 1885. When the Francklyn smelter closed in 1885 August Raht went to Wickes, Montana, and later to Helena, building the plant of the Helena & Livingston Reduction Company, which was consolidated with the Great Falls plant built by Anton Eilers, forming the United States Smelting and Refining Company, of which Raht was manager. About 1889 Meyer Guggenheim and Sons built and put into operation the Philadelphia Smelting Works at Pueblo, Colorado, their chief purpose being to smelt more advantageously than others smelters, the zinky lead sulphide ore of the A. Y. & Minnie mines at Leadville, in which the Guggenheims were interested. The Philadelphia plant was put in operation with metallurgical and commercial results that looked disastrous and would probably have wrecked a concern less strong financially. After a series of vicissitudes the Guggenheims made up their minds that what they needed was the best metallurgical advice and direction possible. Seeking such assistance, they engaged Raht, who took up the work at Pueblo in late 1891 or early 1892. He soon succeeded in putting the enterprise upon its feet and thereby convinced the Guggenheims of the correctness of the principle that the best technical knowledge — no matter what salary or fee may be required to get it — is the cheapest thing in the operation of such a business. Raht continued with the Guggenheims in an operating and consulting capacity for a number of years while they enlarged and developed their metallurgical plants, acting for them in Colorado, Mexico, Perth Amboy, New Jersey, and other places. He visited Australia for the London Exploration Company about 1899 and on returning to the United States, was again connected with the Guggenheim interests, especially the American Smelting and Refining Company, until his retirement from active work about 1910. August Raht died at the Palace Hotel in San Francisco, December 25, 1916. *Engineering and Mining Journal*, January 20, 1917, p. 164; February 24, 1917, p. 358; March 10, 1917, pp. 429-30.

63. *Tribune*, January 1, 1889; January 1, 1894; January 1, 1895, p. 28; January 1, 1896, p. 21; January 1, 1897, p. 29; December 31, 1900, p. 28; *Mining and Scientific Press*, January 18, p. 41; June 28, 1890, p. 431.

64. A. Hanauer, "the eldest of our ore and bullion men," was an active member of the Morgan Company. He took over the Morgan Smelter in 1881. *Tribune*, January 1, 1882.

65. *Tribune*, January 1, 1881; January 1, 1882; June 3, 1888, p. 3; January 1, 1892, p. 22; January 1, 1893, p. 27.

66. *Ibid.*, August 13, 1882.

67. Lynn R. Bailey, *Old Reliable, A History of Bingham Canyon, Utah*, pp. 27-33

68. Gary B. Hanson, "Industry of Destiny: Copper in Utah," *Utah Historical Quarterly*, Vol. 31, No. 3, Summer 1963, p. 271.

69. ASARCO was never a trust. When it was organized there were ten large independent lead smelters, including the Philadelphia, Tacoma, and Selby, and a considerable group of outside refineries. Isaac F. Marcosson, *Metal Magic; the Story of the American Smelting and Refining Company*. New York: Farrar, Strauss, 1949, p. 66.

70. *Ibid.*, pp. 63-69.

71. *Tribune*, December 31 1899, p. 17; Hanson, " Industry of Destiny," p. 271.

72. *Engineering and Mining Journal*, August 8, 1914, pp. 248-253.

CHAPTER FIVE

THE BEEHIVE CHARCOAL KILN, MONUMENT TO BASE METAL SMELTING

Charcoal was the universal fuel prior to establishment of the Pennsylvania coking trade in the early 1870s. Worldwide, charcoal fueled furnaces that produced pig iron, lead, and copper. As late as the 1890s the charcoal iron trade was alive in Wisconsin, Michigan, Pennsylvania, and New York. In the Far West, before opening of coal mines in Colorado, Wyoming, Utah, and Washington, charcoal was the fuel used in restaurants, laundries, assay shops, and homes. Tin and sheet metal shops, and hotels consumed large amounts of the fuel. Until the late 1870s charcoal fueled the base-metal smelting industry from the Pacific Coast to the Rocky Mountains.

Before advent of the masonry kiln (about 1872 in the Far West) charcoal was prepared by charring wood in *clamps* and *meilers*. Both were ancient, fairly simple methods, having been practiced by the Romans for the manufacture of Norican iron, and probably long before that. In the clamp method charcoal was prepared in narrow heaps up to 50 feet long; cord wood being laid within a casing of planks, slabs, or split cord wood on a level or slightly sloping burning ground. Cord wood could be placed in one or two horizontal rows. If one row was used timber length was about nine feet; if two rows were used the wood was four feet in length. The logs were piled, beginning from the upper part, to within a few inches of the top of the casing, then covered with chips, twigs, and leaves, and finally with sand or dirt, filled in against the casing to protect it from fire.

After this was done, the clamp was lite at the lower end, and some of the dirt removed from the upper end to make a draft. Draft-holes were opened in the sides of the casing. As the smoke issued from the holes, more dirt was thrown on, and other draft-holes opened with caution. In this manner the fire was led on till the heat had charred the

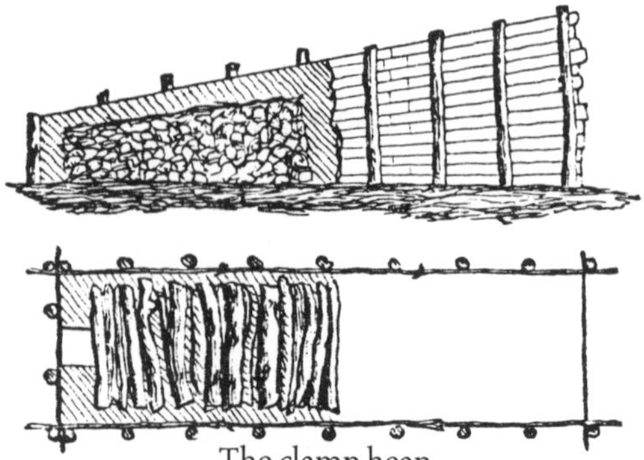

The clamp heap

The standing meiler

The lying meiler with center-block

Before advent of the kiln, charcoal was burned in heaps or meilers, often referred to as pits. Preparation of a meiler varied in the way wood was laid. In the clamp (top illustration) wood was placed between a wood frame. In the standing meiler (middle illustration), wood was laid on end. In the lying meiler (bottom illustration), two, sometimes three tiers of billets, were raised one above the other, and covered with short small wood. To facilitate kindling, a perpendicular block seven to eight inches thick and three feet shorter than the meiler billets, was placed in the meiler. Kindling was done in the three foot deep drum over the center block.

whole. The advantage of this method was that, with a clamp of 50 feet, charcoal could be drawn from the lower end after the fire had progressed about 10 feet, which ordinarily was in 24 hours. If well packed, a clamp of 50 by 9 feet, 6 feet high at the head and 3 feet at the foot, held 15 cords.

The more common method was to burn charcoal in meilers or circular mounds or heaps, from 40 to 50 feet in diameter. In the literature of the American West meilers have universally been labeled charcoal pits, a misnomer. Other than carefully preparing the ground by leveling, compacting, and drying, there was no excavation involved in construction of meilers.

There were two kinds of meilers: lying and standing. In the lying meiler wood is laid on its side, in the standing meiler sticks of wood were set close together in a nearly vertical position. In the center of both large sticks were placed sufficiently wide apart to form a chimney. After the wood was carefully arranged, brushwood and loose earth were thrown over the pile to smother the flame and prevent its bursting out from the mass of wood. The fire was applied to the wood soon after this covering was effected, which was increased from time to time as circumstances required. For the purpose of attracting the fire all around the heap, holes were made in the sides to create draft through which the watery elements of the wood were expelled. Were the air allowed to circulate, the entire mass of wood would be reduced to ashes. The burning lasted two or three days and nights, according to the amount and nature of the wood and the success attending the operation. Success of the process depended on the skill and watchfulness of the colliers, as well as the state of weather.

Charcoal burning on a commercial scale was introduced to the Far West simultaneous from Utah and California. In the latter state vast stands of timber on the eastern scarp of the Sierra Nevada were utilized by mines and mills of the Comstock, then by the smelters at Cerro Gordo, and later by Nevada's argentiferous lead districts. The Truckee firm of Sisson, Wallace & Company, which advertised as a "Charcoal Agency," cornered the Sierra Nevada charcoal trade, selling the fuel as far east as Salt Lake Valley for between 25 and 28 cents a bushel. According to company ads, the trade was seasonal, "orders . . . received and arrangements made by June 15 . . . would receive immediate attention. Quantity was no issue." That was true, for great quantities of Truckee timber was harvested and burned, the charcoal bagged and shipped in wagons or over the transcontinental railroad in specially prepared racks that lessened jarring.[1]

The foundation of Sisson, Wallace & Company's business, as well as other fuel suppliers, was immigrant labor, mostly Italians and Swiss. It is not known how they were recruited. It may have been through provisions of the Contract Labor Law, which allowed employers to bring foreign workers into the United under an agreement based on a maximum of 12 months labor for passage. Or they may have drifted into the Far West from the Great Lakes charcoal iron region, and a few may have come from the Deep South where charcoal burning was also widely practiced. Regardless of where they came from, they were known as *carbonari*, men who had learned charcoal burning in their native land.

Production of charcoal had a long history in Europe, rising to prominence in England, Sweden, Germany, France, and Italy during the industrialization of the late 18th and early 19th centuries. A charcoal burners guild known as the *carbonari* developed in France in the late 1770s and was introduced into Naples in 1806, and spread throughout Italy after the Napoleonic period. The carbonari were not only skilled workers, they were activists who espoused the cause of Italian unity. It was not long before carbonarism became synonomous with the ideas of independence, unity and liberty. Although they failed to reach that objective, the carbonari who emigrated to America were a free-thinking, independent lot, who introduced into the Great Basin the methods of charcoal burning in meilers. For nearly two decades they kept the furnaces going at Eureka, Nevada.

In some respects Eureka was a charcoal burners paradise. By the mid-1870s the five furnaces of the Eureka Consolidated Company were consuming 4,600 bushels daily, and the Richmond Consolidated Company a similar amount. The carbonari settled around Eureka, laid claim by squatter rights to tracts of cedar, mountain mahogany, and nut pine, and built rough hewn cabins near their meilers. Such settlements were called charcoal ranches.

Mining companies did not deal directly with the carbonari, however. Instead the smelters received the fuel from teamsters who picked it up at the burning grounds and hauled it to Eureka. The fuel, which initially was priced at 15 cents a bushel, was weighed at the smelters and a receipt presented to the teamster verifying the amount of charcoal received. The receipts were then turned over to charcoal burners who received payment, usually in goods or credit, at some mercantile company. It was a system prone to fraud.

Nevertheless, the population of carbonari grew in Nevada's argentiferous lead districts. In 1870 the number of Italians and Swiss in the State numbered less than 400. Five years later Eureka County alone had about that number. By 1878 there were an estimated 2,000 charcoal burners at work producing fuel for Eureka's smelters. Their production was phenomenal and devastating. In spring of 1874 the *Eureka Sentinel* commented that "the available fuel supply is every day becoming scantier and scantier under the increasing demands which our furnaces have each successive year made upon it. . . ."[2]

With its furnaces consuming nearly four million bushels of charcoal a year, Eureka faced a charcoal shortage of giant proportion by 1874. Completion of the Eureka & Palisade Railroad partially alleviated the crisis, permitting the tapping of Pine Nut Valley, 35 miles southeast of Virginia City, in western Nevada.[3] With miles of pinyon, nut pine, and mountain mahogany, Pine Nut Valley was an immense fuel resource. Because snowfall was light in the valley, charcoal burning went on year round. On average 80 men were kept constantly at work, with less men employed during the summer. In summer of 1877 the valley produced 150,000 to 200,000 bushels of charcoal, most of which went to the Eureka Consolidated and Richmond companies.[4] Eureka's fuel shortfall was made up by Utah, the second center of charcoal burning.

* * *

During the infancy of the smelting industry charcoal was burned in the Wasatch Mountains and in piñon districts to the west and south. Although Utah charcoal was several cents cheaper than the California product, it was inferior in quality, and the supply was "precarious." Because the meiler process of charcoal production was affected by Utah's winters it was impossible for entrepreneurs to make money in the business. That was changed by James C. Cameron, Jr., a mining engineer with extensive experience in the production of charcoal iron.[5]

James C. Cameron, Jr., was a scion of a distinguished family that participated in Pennsylvania's iron trade and railroad transportation. His grandfather, Simon Cameron, planted the seed of this sturdy Scotch family in Donegal Township, Pennsylvania, in 1777. Simon later moved to Maytown, Lancaster County, thence to Sunbury and

Lewisburg, where he raised his sons: William (born October 15, 1795), Simon (born March 8, 1799), and James (born March 1, 1801). All of the Cameron boys started out in life without pecuniary assistance. Becoming tradesmen they worked their way to distinction. In 1816 Simon Cameron apprenticed himself to Andrew Kennedy, publisher of the *Northumberland Gazette*. The printing trade took Simon to Harrisburg in 1820, where he worked on the *Pennsylavina Republican;* later he worked on the *Congressional Globe*.

In his youth James learned the blacksmithing trade and then followed in his brother's footsteps, becoming a printer and editor. For a time he was an owner of the *Lycoming Gazette*, and in 1829 obtained control of the *Lancaster Sentinel*. He read law with James Buchanan, who became president of the United States, and with Judge Molton C. Rogers, of Lancaster. James was admitted to the bar in 1829, served one term as district attorney, and practiced in Lancaster City until 1846. He was admitted to the bar of Northumberland County on August 4, 1851.

In 1837 Simon, James, and William joined Thomas McNair and Samuel Headley in the iron trade, purchasing an iron bloomery called Foundryville or Columbia Furnace, about two and a half miles north of Berwick. Foundryville was a typical Pennsylvania iron plantation, consisting of a cupola furnace, foundry, machine, wheelwright and blacksmith shops, as well as a store, dwelling, and other buildings. A large acreage (2,400 acres) of woodland furnished material for making charcoal. The Columbia Furnace did not own mineral land, but procured iron ore under contract from miners at Bloomsburg.

The Camerons and their partners produced pig iron until at least 1846, much of their product being made into castings, hollow ware, and stoves. The manufacture of car wheels and turnouts (called frogs) for the Harrisburg and Lancaster Railroad, became a large part of their business. This propelled both Simon and James into the ranks of railroad officialdom.

In 1845 Simon Cameron campaigned for the office of United States senator from Pennsylvania, to which he was elected. He served on the board of directors of the North Central Railroad from its incorporation in 1854 until his resignation in April 1862. And he was Secretary of War in Lincoln's cabinet from March 11, 1861, until January 11, 1862. His brother's career fell short.

In 1839 James was appointed superintendent of motive power on the Columbia Railroad. At the same time he was superintendent of the Harrisburg and Lancaster Railroad, and later became superintendent of motive power of the Philadelphia and Columbia Railroad, one of Foundryville's best customers. In 1843 he was appointed deputy attorney general of the Mayor's Court in Lancaster, in place of Samuel Humes Porter, who resigned. During the Mexican War James was sutler at New Orleans; in fall of 1848 he went to Gosport, Virginia, where he remained until 1851 engaged in building the steamship *Powhatan* and repairing other vessels for the government. For some years he operated several farms near Milton. When the Civil War broke out he was stationed at Sunbury as superintendent of the Northern Central Railroad. Through his brother's cabinet connections, James was commissioned colonel of the 79th Highlanders regiment of New York and was killed in the battle of Bull Run, July 21, 1861. His brother, Simon, died a millionaire at age 90 years.[6] Such was the Cameron family.

Born about 1830, James C. Cameron, Jr., was indoctrinated by his father into iron smelting at Foundryville. During the Civil War he continued in the iron trade in Lancaster County. It is not known if James attended college but somewhere along the way he attained the equivalent of a mining engineering degree. In the mid-1860s he went to the Lake Superior iron region, where he built charcoal kilns for the Central Iron Works at Lauton in Marquette County, Michigan.

The Michigan environment caused Cameron to reassess what he had learned about charcoal burning at Foundryville. In Pennsylvania charcoal was burned in heaps or meilers.[7] But there was a big difference in weather between Lancaster, Pennsylvania, and Marquette County, Michigan. The Lake Superior region was subject to heavy snowfall. Storms sweeping in from the Great Lakes often curtailed charcoal burning, shutting down Michigan's iron industry for prolonged periods of time. To keep the furnaces operating, fuel was manufactured in enclosed masonry kilns, less affected by inclement weather. Because James C. Cameron, Jr., would play a large role in construction of masonry charcoal kilns, the earliest types need to be looked at.

Masonry charcoal kilns were of two shapes prior to 1868: rectangular and round. First appearing in Sweden in the early 19th Century, rectangular kilns were constructed for a number of reasons: to avoid keeping large stores of charcoal, to obtain it dry and fresh, to save labor, and overcome the faults of inferior workmenship.

Generally speaking, the advantage of kilns over heaps was not great. Charcoal made by both methods had identical strength; the only real advantage being that kilns could be operated in any kind of weather. Both rectangular and circular kilns were usually built of red brick, or rarely, of brick and stone together. Occasionally, refractory brick was used, but that was not necessary. Foundations were stone. Several precautions were followed in constructing the walls. Bricks were tested before laying up the walls to asssure they were sufficiently hard to resist heat. Second or third quality firebrick was not used.

As pyroligneous acid from distillation of wood attacks lime mortar, a fire-clay mortar, to which a little salt was added, was used to lay up the brick. Loam mixed with coal-tar and salt was also used. Whatever was utilized to fill the joints, special care was taken in laying the bricks to assure every joint was filled so that no air would be admitted, materially decreasing the yield of the kiln. Frequently headers were put in to tie the bricks together. The floor of early kilns were made of two rows of brick set edgewise, carefully laid; later floors were made of clay, or any material that packed hard. The kiln's center was about six inches higher than the sides, which were brought up to the bottom of the lower vents. Most kilns were carefully pointed, and then painted on the outside with a wash of clay, and covered with a coating of coal-tar, which waterproofed them for several years. Some kilns were roofed with rough boards to protect the masonry from weather, a practice that was discontinued early as having no special advantage.

The earliest rectangular kilns varied in size from 40 to 50 feet in length, and 12 to 17 feet in width, and were 12 to 18 feet in height. Capacity varied from 30 to 90 cords of wood. Rectangular kilns at the Michigan Central Iron Works were 48 feet long, 17 wide, and 18 in height, and had a yield of 4,000 bushels of charcoal.

Vertical walls of whatever height were 1½ brick thick, containing from 20 to 22 bricks to the cubic foot of wall. To insure sufficient strength to resist expansion and contraction due to heating and cooling, they were frequently provided with buttresses one brick thick and two wide. (At Wassaic, New York, kilns were buttressed; those at Lauton were not.) Walls were supported with strong braces, three to four feet apart, made of round or hewn wood, or of cast iron, which were buried in the ground below, and tied above and below with iron rods, the lower end passing beneath the floor of the kiln. When made of wood they were usually eight inches square, or round, placed edgewise. They were sometimes tied at the top with wooden braces of the same size, securely fastened by iron rods running through the corners.

When a number of kilns were built together, as at the Michigan Central Iron Works, only the end kilns were braced, the intermediate ones being supported below by wooden braces, securely fastened at the bottom. The roofs were arched, one brick, or eight inches thick, laid in headers, 14 being used in each superficial foot. Many of the kilns had a round hole in the center, from 16 to 18 inches in diameter, which was closed by a cast-iron plate. A kiln of 45-cord capacity required from 35,000 to 40,000 bricks; a 90-cord kiln from 60,000 to 65,000.[8]

Rectangular kilns had two doors at one end, one above the other, both used for introduction of wood. Discharge of charcoal was made by the lower door only. These doors were generally five feet square, made of cast iron in two parts, supported by hinges on a cast-iron frame. Doors were sometimes made of a single piece of sheet iron. On the sides between the braces or buttresses, rows of openings called vent-holes, were regularly spaced, vertically and horizontally, for introduction of air necessary for combustion, and to allow distillates to escape. Vertically there were three or four rows. Between the braces, horizontally, there may be only one row where the braces are very near to each other; or two rows for the lower and one for the upper, or three in each row. The number was regulated by the manner of bracing. The rows were always placed one directly over the other. To each row a name was given: the lowest row was called the foot vent; the middle one the knee vent, and the upper one the shoulder vent. On the ends of the kiln the vents were placed less regularly than on the sides. Sometimes there were only one row, one above the other; sometimes one row not in line, and sometimes there were more. From three to four vent-holes were usually made in the lower doors. These openings were usually of a size to admit a single brick.

The wood was usually cut to a length of four feet, laid flat on the

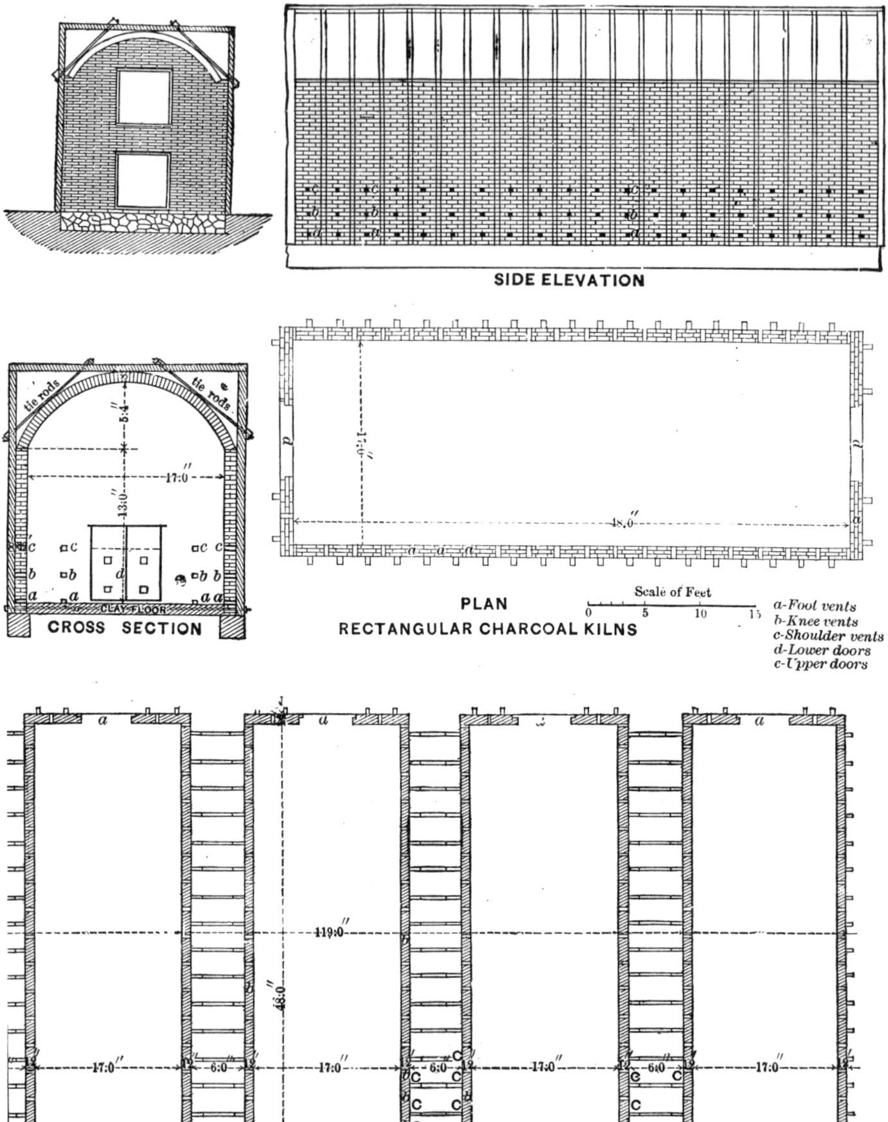

Rectangular charcoal kiln.

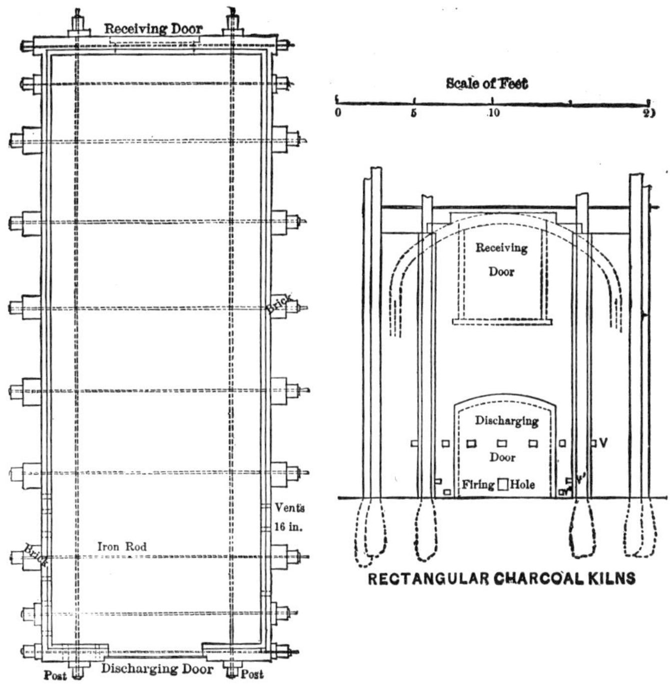

RECTANGULAR CHARCOAL KILNS

floor in the direction of the length of the kiln, and piled as closely as possible to leave few interstices. The sticks were piled in successive steps, so as to fill the kiln to the arch. It took four men, working one day, to fill a 45-cord kiln.

Several methods were used to light the kiln. When first used in New England, kilns were lighted in one corner, the fire being drawn along the long side, and round the kiln on the outside of the wood, so that heat acted directly on the walls. The preferred method of lighting was from above through a chimney left in the wood in the center of the kiln, exactly as in the meiler, the fire being drawn down by the openings in the sides; this chimney may or may not connect with a channel leading to the discharging door. This method, used in the American South and Southwest, and Mexico, was preferred when kilns were very wide.

Sometimes kilns were lite by means of a channel built through the middle of the kiln, having an opening at each door. Filled with dry wood and shavings, it was lite from the back door, the fire being drawn through the wood to the front door by manipulating the vents. When the fire reached the front both doors were closed, the whole kiln being consided lite. This method was called the center burn. As the fire was generated in the wood in both methods, the heat did not affect kiln walls.

Success of the process depended on closing vents at the proper time. As soon as the kiln was well lite the two lower rows were closed. Vents of the upper row were closed little by little. Those of the middle in the upper row being closed first; then those from which blue smoke came in their turn. When vents of the first row were being closed, those of the middle were opened, and so on. One man by day and one man by night easily superintended five or six kilns. They had little to do except to draw the fire regularly down, to watch the color of the smoke so as to close vents at the proper time, and to close any cracks that may form in the kiln. The charge was ready to be drawn from the kiln if after two or three days there was no flame or smoke issuing from a few vents that had been opened. If the kiln was opened too soon water had to be pumped in at once, as the charcoal would ignite.

The rectangular kiln went out of favor because there was so much space to heat. Its thin walls required a large number of braces to secure a tight kiln. Once cracked, the rectangular kiln was difficult, if not impossible, to repair and make tight again.

Round kilns differed from rectangular ones only in their shape. Arches of these kilns were supported by iron rings at their base. They had three rows of vents, one at the base, with six courses of brick between each of the others. Vents were two by four inches and 18 inches apart. Round kilns were usually 28 to 30 feet in diameter at the base, and 26 to 28 feet at the spring of the arch. Vertical walls were 11 to 12 feet to the spring of the arch, and one foot thick. The arch was eight inches thick, laid in headers. Such a kiln had about 300 cast-iron vents weighing about eight pounds each. Thus a round kiln of 40- to 45-cords capacity required more precautions in building when the walls were not battered, the usual braces being hooped with strong iron bands. The doors were made of No. 10 sheet iron. The top opening was a cast-iron ring, which together with the lintel of the door weighed about 1,500 pounds. Wrought-iron bands around the kiln weighed about the same. These kilns, requiring about 36,000 bricks, were usually built against a bank so that the top could easily be reached.

Filling was commenced by placing pieces of wood about five inches in diameter, leaving interstices for air-channels; above this about 18 inches of dry wood and brands were placed, leaving a space

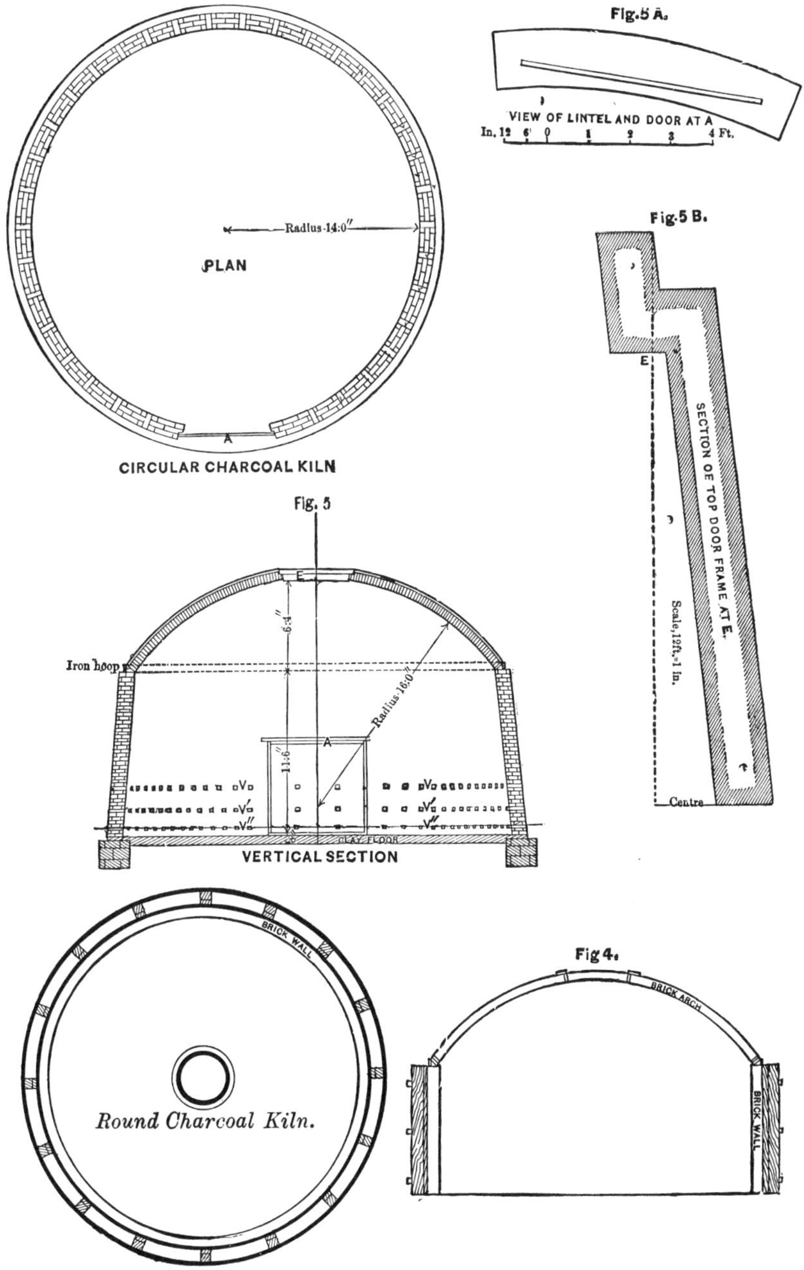

Round charcoal kilns.

in the center about four feet in diameter for a chimney, which was carried to the top of the kiln. The wood was piled horizontally and packed tight. With ordinary facilities, it took four to five men one day to fill a 45-cord kiln with wood four feet long. Where butt-ends of logs were used, it took 12 hours to fill a 35-cord kiln; often tumbled in, such wood did not require much handling. It took five to six days after all vents were closed to cool a kiln of from 35 to 50 cords. So long as there was heat in the kiln it could be felt on the sheet iron doors. As soon as these were cold, the kiln was cool enough to draw.

Both rectangular and round kilns were unstable. Rectangular kilns required wood and iron bracing. Circular kilns, which offer no advantages over the rectangular, besides being braced, had to be hooped with iron. Thus the structures were expensive, and did not last long, contraction and expansion cracking the walls, which could not be effectively closed.

Such were the problems when James C. Cameron, Jr., arrived at the Central Iron Works. He went to work to reduce the expense involved in building, maintaining and operating rectangular and circular kilns. It did not take him long to iron out the company's fuel problems. About 1867 or '68 he designed an entirely new kiln.

Cameron's charcoal kiln, first built of red brick, was conical shaped and smaller than the two previous types. Being from 20 to 25 feet in height and from 25 to 30 feet in diameter, it was intended for 25 to 45 cords of wood, depending on size. The walls were 12 inches thick at the bottom, but at a height of seven feet four inches they diminished to eight inches. From that height upward all bricks were laid in headers. At the height of 23 feet the kiln was capped by another cone, five feet four inches high, and of much flatter angle. On one side a doorway, seven feet high and six feet wide, was built, into which a sheet-iron door was fitted against flanges made for the purpose. For charging the kiln a second door of the same dimension, was built immediately under the top of the kiln. These doors had iron frames.[9]

It took 33,000 to 40,000 bricks to construct a conical kiln. The floor of the kiln was made of tempered clay eight to 12 inches thick, well beaten with mauls. It was very nearly level, being four inches higher in the center than at the sides, and came up to the bottom of the lower tier of vents. The kiln had three rows of vents.

Whether built of brick or stone, the conical kiln needed to be

Fig.7.

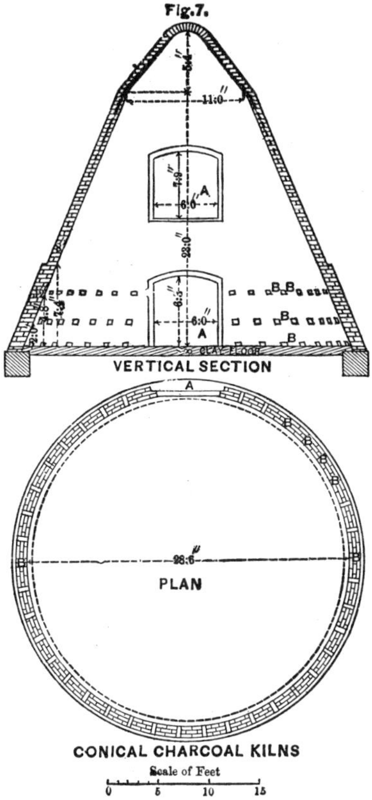

Fig 8

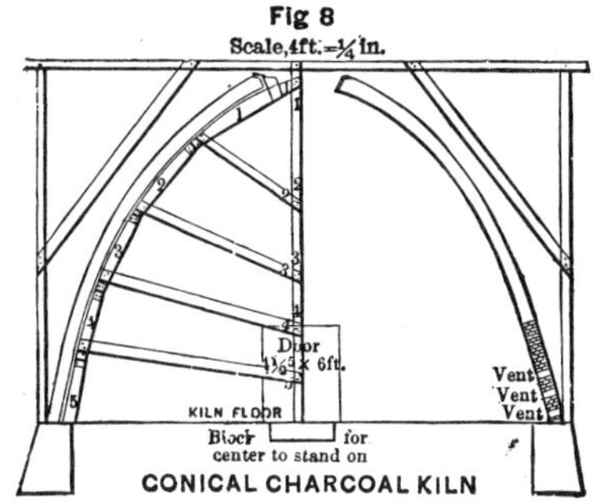

Fig.6

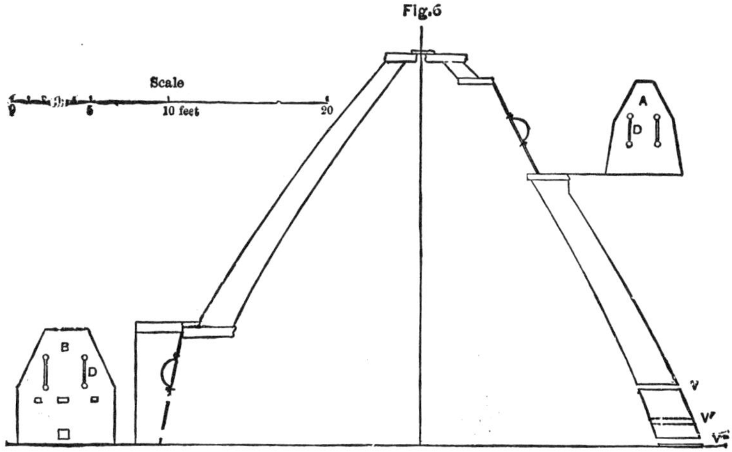

Conical or beehive charcoal kiln as designed by James C. Cameron, Jr.

Loading charcoal kilns along the Fox River north of De Pere, Wisconsin. (Courtesy Wisconsin Historical Society)

strong enough to resist the low heat of operation. Constructed in this way the kiln required no braces of any kind. Placement of the Cameron kiln was an important factor in its economical running. A hillside location was preferred, a part of the earth being removed so that the ground was flat in front of the discharge door. On the opposite side of the kiln, ramps were extended from the hillside to the loading door at the top of the kiln to allow loading of cord wood by means of wheel barrows.

A 35-cord kiln yielded 1,750 bushels of charcoal. Both soft and hard wood was burned, although the former was generally used in the manufacture of blooms. To charge the kiln, skids two inches in diameter were laid three feet apart, keeping the ends up to the outside. These had to be placed in the radii of the circle. Wood in all sizes from four inches to two feet in diameter was cut into lengths of four feet eight inches. A fireplace or chimney four feet square was made in the center of the kiln, and carefully preserved in the filling to the top. An air-channel was made from it to the opening in the lower door. This was filled with shavings, brands and light wood, thrown in loose, which was lite with a long pole from the door. Eight men could look after 13 kilns. It took 13 days to fill, burn and empty a 35-cord kiln (one day to fill and one to empty). Thirteen kilns could make 22 turns a year.

The conical kiln was easily and cheaply built. In Michigan where brick cost $17.50 per thousand, such a kiln could be built for no more than $600. Repairs posed no problems. Its small, compact size allowed for rapid loading and discharge. And most important, a few men could tend a large number of kilns. M. R. Hunt, Secretary of the National Iron Company of Wisconsin, endorsed the Cameron kiln, reporting that it was "superior to any charcoal kiln now in use, affording a saving in construction over the rectangular from 50 to 75 per cent; in cost of manufacture 25 to 50 percent, and a very favorable profit in quality and quantity of coal produced, while over the pit [heap] process they possess very decided advantage. During the winter season, when snow abounds, they can be worked with almost as great economy as during the summer months."[10]

Because it solved many problems of the fuel trade, Cameron's "beehive" kiln, as it was called, was immediately accepted in charcoal iron districts. Batteries of conical kilns were built throughout Michigan and Wisconsin, and the design was carried to Norton's Iron Works at Plattsburg, New York. From there the conical kiln spread to Wassaic, Dutchess County, New York, and Readsboro, Vermont. In Utah there were mining men who thought Cameron's kilns would meet their fuel needs as well.

Among those men were the owners of the Miller Mining Company in American Fork Canyon: the Vanderbilts of New York, Lloyd Aspinwall, L. A. Hardins of Chicago, and R. N. Baskin of Salt Lake City. They contracted with James C. Cameron, Jr., to build 15 conical kilns to supply charcoal for their Sultana Smelter. In spring of 1871 Cameron arrived in Utah with Hugh McAllister, an expert in charcoal burning from the National Iron Company of Wisconsin. Within weeks of his appearance on the Utah scene, Cameron had assigned inventor's rights to Evans and Morris, Salt Lake City manufacturers of firebrick, to build kilns in Utah, Nevada, and California.[11]

Fifteen conical kilns were erected against a mountainside within a few hundred feet of the Sultana furnaces. The overall design was for smooth, economical production of charcoal. A sawmill 200 feet higher on the mountain processed pine and aspen cut from Wasatch groves, the cord wood fed to kiln loading doors. Three more sawmills were added before end of 1871. A tramway, supported by trestles, connected the discharge doors of the kilns to the smelter feed floor so that charcoal could be dumped at the mouth of the furnaces. In that way the fragile fuel was handled only once. Sheds were constructed to store excess charcoal.

Being 26 feet in diameter at the base, 20 feet in height, each kiln had a capacity of 25 cords. It took nine to ten days to burn a kiln. At end of five days all vents were closed, and the kiln was no longer watched. The fire was extinguished, with water if necessary, and the charcoal taken out of the kilns. After every burning the kilns were whitewashed to seal cracks.

Since each kiln was capable of producing 1,200 bushels of charcoal every ten days, or 45,000 bushels a month, these kilns could supply the three furnaces of the Sultana Works without interruption,[12] and still have charcoal left over. After discounting construction and labor costs, the Miller Mining Company produced charcoal at 12¾ cents per bushel.[13]

Considering this was one-half the price of fuel paid by other Utah smelters, the Miller Mining Company moved to supply charcoal to smelters in Salt Lake Valley. In 1874, upon completion of the American Fork Railroad, the Miller Company shipped about 25,000 bushels monthly to all parts of the valley. As a *Tribune* correspondent commented "it doesn't require a man with a glass eye to . . . see that it therefore gives a good big squad of men employment. . . ." The American Fork Railroad sent 10 to 15 cars down the canyon every day, loaded with charcoal, bullion, ore, and lumber.[14]

As far as it could be, the Miller Company's charcoal operation was mechanized. Fuel destined for smelters in Salt Lake Valley was transferred from the kilns to nearby sheds, where it was thrown upon tables three feet high and four feet square. In the middle of each table was an 18 inch square opening to which a bag was hooked capable of holding two to three bushels of charcoal. The men pushed the charcoal into the bags. It took one day for six men to empty a kiln, sack the charcoal, and put the bags onto cars for transport down the canyon.

American Fork Canyon charcoal was made from pine and aspen, soft woods that produced soft charcoal. Because it could be crushed by its own weight, this commodity did not stand transportation over long distances. To protect it, bags of charcoal were shipped without much packing in specially constructed cars. Nevertheless much fine coal was lost. Use of soft woods to manufacture charcoal was justified by the fact that the only alternate fuel, coke from Connellsville, Pennsylvania, cost $45 per ton in Utah. Regardless of some fuel loss, by 1873 the American Fork Canyon kilns were "in every particular a success."[15]

James C. Cameron's introduction of the conical kiln into Utah, and its successful trial in American Fork Canyon, prompted others to enter the charcoal trade. In late 1873 or early 1874 the Hilliard Flume and Lumber Company was organized in Salt Lake City by John W. Kerr, a smelterman, William K. Sloan, a Salt Lake City banker, Henry Wadsworth of Wells, Fargo & Company, and Fred Myers of the Galena Smelter. Their field of operation was on the northern slope of the Uinta Mountains in Summit County, nearly on the line between Utah and Wyoming territories. Their objective was three fold: to supply the Union Pacific Railroad Company with ties, the towns along its line and Salt Lake City with lumber, and to furnish Utah's smelters with charcoal.

Their scheme was of stupendous proportions, involving an outlay of millions dollars before it produced any returns. The risk was worth it. At that time there was a growing demand for lumber and charcoal, with a prospect of steady and ever increasing growth. In fact, these men believed there was a market for all they could produce. And they had access to extensive forests within 30 miles of the Union Pacific Railroad, "a vast field of underdeveloped resources," which they moved quickly to exploit.

A large steam sawmill was erected in the forest on the headwaters of Bear River. In July 1874, 700 men set to work falling trees and sawing lumber to construct the longest flume in western America.[16] In one hundred days a 27½ mile long flume was put together. Its construction called for 2,500,000 feet of lumber and 5,000,000 feet of timbers which supported the flume. A hundred tons of spikes and nails held it together, the whole costing $2,000,000.

The V-shaped flume, made of two-inch lumber, was four feet across at the top. In some places in its long and almost straight course to the railroad, it hugged the ground; where it crossed ravines and rivers, it was supported on high trestle work. Its fall varied from 100 to 300 feet to the mile, so that when the flume was nearly full of water, lumber thrown into it descended with astonishing velocity, the time required for sawed logs to reach Hilliard being two hours and fifteen minutes, and for unsawed pieces, two and a half hours. Some notion of the transporting capacity of this contrivance may be gathered from the fact that in 40 days 18,000 cords of wood were landed at the railroad, where it was piled up 20 feet high for over a mile.

The head of the flume has three branches, each running into a different portion of the forest on the northern slope of the Uintah Mountains. There the company had its steam sawmill, which turned out daily 35,000 to 40,000 feet of white pine lumber which was almost as free from knots as that produced at Truckee, and of a more durable character. The lumber, run down the flume during the summer, was sold and shipped east and west by rail. Cord wood, however, was converted to charcoal at a point on the Union Pacific Railroad about 100 miles east of Salt Lake City that would be christened Hilliard, where with James Cameron's blessings, the company built 29 conical kilns, at a cost of $800 each.

In some ways the Hilliard kilns were similar to those in American Fork Canyon, being 26 feet in diameter and 21 feet in height, with rock foundations and door arches. The walls, however, were of adobe bricks, varying in thickness from 15 to 25 inches. Ten kilns had a capacity of 2,500 bushels each, the remaining 19 of 1,800 bushels each. With a burning time of about ten days, the same as American Fork Canyon, the Hilliard kilns had a combined capacity of about 2,000,000 bushels a year. The Hilliard Company started out in fall 1874 shipping 80,000 bushels of charcoal monthly, 60,000 of which went to the Flagstaff smelter.

A year later (in 1875) the company secured special freight rates on ore shipped over the Utah Central and Union Pacific railroads and erected a one stack smelter at Hilliard, adjoining its kilns. Although small, the smelter was "quite complete, having all the necessary machinery, including a Dodge crusher, a 40-horsepower steam engine and an improved blower," so that additional stacks might be put in when business warranted it. In February they commenced reducing ore from the Carbonate mines on Kesler's Peak in Big Cottonwood, which the Hilliard Company owned. The Old Telegraph and Mayflower mines in Bingham Canyon also contributed ore.

In spring of 1876 a second furnace — a giant water-jacket 10 feet high by six feet in diameter — was constructed. By July it was working "like a charm," producing 15 tons of bullion every 24 hours. Both blast furnaces had a common dust chamber on the roof of the building, which collected the dust and deposited it on the feed floor near the furnaces. Like the Miller Company in American Fork Canyon, the Hilliard smelter and charcoal kilns were designed as integrated units.

Charcoal was run in a light car from the kilns to the roof of the smelter building where it was dumped into bins on the feed floor. A rail track was laid into the building so ore, flux, charcoal, and bullion could be weighed before it was received or shipped. There was also an assay office which allowed the company to do its own sampling.

The first Hilliard furnace produced only a few tons of bullion. But addition of the second, "giant," water-jacket furnace transformed production. Together, the furnaces reduced 3,888 tons of ore in 1875, producing 1,187 tons of bullion, carrying 150 ounces of silver and about an ounce of gold to the ton. In full production, the furnaces consumed 20,000 bushels of charcoal a month.

In connection with its smelter and kilns, the Hilliard Flume and Lumber Company maintained a blacksmith shop and a brick kiln. A general merchandise store was established to accommodate its employees, about $50,000 worth of goods passing over its counter each year. A telegraph line connected Hilliard to the head of the flume; "a convenience which was absolutely necessary to the complete management of the business."

Hilliard was not a large community, only 30 families. About 25 men were employed at the smelter, 15 at the kilns; enough to support a saloon or two.[17] And there was sufficient traffic through town for a hotel and several boardinghouses. The town had a free school housed in a well furnished log house built by the company, supported by a fund of $1,200 a year raised by a special tax voted by the people of the school district. The Hilliard school enjoyed the reputation of being the "best in Uintah County." Although Hilliard's kilns and smelter were in Wyoming Territory, Utah claimed it "because the ore smelted there was from its mines and the lumber and fuel produced there were from the woods of Zion."

The principal market for Hilliard charcoal in 1874 was Salt Lake Valley, price of the commodity delivered at Sandy ranging between 23 and 25 cents per bushel. Limited quantities were also shipped as far east as Omaha, and westward to Eureka, Nevada. Business increased substantially over the next few years until 1877 when the Hilliard Flume and Lumber Company disposed of 1,250,000 feet of lumber, and nearly a million bushels of charcoal. Over the next four years this company shipped enormous quantities of lumber and charcoal. In late 1881, however, the Utah-Wyoming Improvement Company, composed of Boston capitalists, purchased the assets of the Hilliard Flume and Lumber Company and shipped lumber and charcoal a few years longer.[18]

Success of the Hilliard Flume and Lumber Company and the Miller Mining Company brought smaller operators into the charcoal trade. E. L. Pease, owner of a sawmill at Evanston, Wyoming, built three kilns near his mill. And the Wyoming Charcoal Company, also at Evanston, erected six kilns which it "vigorously operated," sending 200,000 bushels of charcoal to market in 1876. Wood cut in the Uinta Mountains and floated down the Bear River supplied both companies.[19]

Furnaces in Nevada had fuel problems since inception of base metal smelting. Ever since the White Pine excitement charcoal had been burned in heaps. The result was an inferior fuel full of stones and dirt. To the west at Ward in White Pine County, the Martin White Company which had two furnaces, seized upon the Cameron kiln as an answer to contaminated fuel. In July 1876 the company brought in a Mr. Morrison from Hilliard, Wyoming Territory. Reputed to have had "a life long experience in the coal business," this gentleman contracted to build six Cameron kilns of a capacity of 800 bushels of charcoal daily. Morrison agreed to furnish all the charcoal required to run the two furnaces at 13 cents per bushel at the kilns. Haulage of charcoal from kilns to furnaces cost two cents per bushel more. The result was a superior fuel for smelting purposes.[20]

Between 1873 and 1881 there was a proliferation of conical kilns from Inyo County, California, to Idaho, and from Wyoming to the Gila River in Arizona Territory.[21] Most, however, were built in Utah between 1872 and 1878. Cameron kilns were erected at the Union Iron Works near Cedar City, in American Fork Canyon, and in Soldier Canyon on the west side of the Oquirrh Mountains. The latter location supplied fuel for the Waterman Smelter at Stockton. As related, a German merchant named Charles Lamendorf, branched into the charcoal business, constructing conical kilns at Tintic.

A "veteran coal burner" by the mid-1870s, Lamendorf headed south to Beaver County. In the Wah-Wah range, 14 miles from Frisco, he established a camp, called Lamendorf, and turned his men loose chopping wood and burning charcoal for smelters of the district. According to the *Tribune* of February 20, 1878, were it not for the "big [charcoal] business here, many of the Saints in this [Beaver County] and the adjoining counties would go hungry." There was truth in that statement, for not only did Lamendorf have coaling sites in the Wah-Wah Mountains, the Frisco Company erected five conical kilns adjacent to their smelter in July 1878, and there were seven more kilns to the east at Copper Canyon. There were considerable more kilns in the San Francisco District than that. The 10th U.S. Census (1880), stated that "thirty-six bee-hive kilns supply the furnaces of the district with fuel. They were in eight groups, under separate individual management, at places where wood was accessible, and at distances of from 6 to 18 miles from Frisco."

According to the Census, all kilns in the San Francisco District were of similar construction:

> made of granite float found in the neighborhood and a lime mortar. Their sizes varied from 16 to 26 feet in diameter; height of the kiln being equal to the diameter. Wall thickness varied from 18 to 30 inches at the base and from 12 to 18 inches at the summit. They were identical to other Cameron

Charcoal kilns at Copper Canyon, seven miles east of Milford, Utah. (Photo by Dexter Oliver)

kilns: each had two openings, closed by sheet iron doors, one at ground level, 4 by 6 feet; the other in the side two-thirds of the distance to the apex, 3 by 4 feet. There were also three rows of vent holes, 3 by 4 inches, near the ground. The lower row is at the surface of the ground. The rows above are 18 inches apart, having vent holes 3 feet apart in each row. The kilns cost from $500 to $1,000 each.

Sixteen-foot kiln held about 15 cords of wood and the 26-foot kiln 45 cords. The wood was all piñon pine, cut at all seasons by Mormons at $1.25 per cord. It was brought from one to four miles by sledges or wagons to the kilns for from $1.50 to $2.50 per cord. The kilns are fired in the center at the bottom (though sometimes at the top), and the fire was drawn to the top by leaving a small unsealed space around the upper door. This is then closed entirely, and the fire is regulated by the vent holes. The duration of burning is from three to seven days, and of cooling from three to six days. Charring, which includes packing the wood in the kiln and drawing the coal, is usually done by contract, and costs from $2^3/_4$ to $3^1/_2$ cents per bushel. About 50 bushels are produced per cord charred. The coal is bought by weight, 17 pounds making a bushel. It is

View of charcoal kiln, discharge side, Milford, Utah. Journalist and photographer Dexter Oliver stands at discharge door.

> shipped to the smelters in racks, at a cost of from 3 to 5½ cents per bushel, depending on the distance. The price received is 18 cents per bushel. Kiln hands are paid from $2 to $2.75. The labor required averages one man per kiln per twenty-four hours.[22]

Thanks to James C. Cameron the manufacture of charcoal became "a home industry" in Utah, at least for a brief period of time. In 1873, 10,797,650 pounds, or 540 carloads of charcoal, was received by the Utah Central Railroad Company. This was equivalent to $20,000 a month being paid out for the commodity by various smelters. In 1874 the railroad carried 16,032,000 pounds; in 1875, 15,756,000 pounds; and in 1878, the amount dropped to 6,420,000 pounds, and remained about the same in 1879. After that date Utah smelters reduced their reliance on charcoal. Tin and sheet metal shops, laundries and hotels continued to use charcoal as fuel. Thereafter demand for the product dropped steadily, forcing many dealers out of the charcoal business. By 1887 charcoal production was largely in the hands of two or three men who had permanent kilns. Sale of charcoal in Utah in that year amounted to only $28,125. Another fuel had replaced it.[23]

* * *

For at least a dozen years charcoal was the fuel of choice. In 1871 it was easily procurable at no more than 22 cents a bushel. Opening of base metal deposits in Nevada and Utah spurred smelting, which in turn increased demand for the fuel. Within a few years annual consumption reached nearly four million bushels at Eureka alone. Since a cord of nut pine wood produced about 28 bushels of coal, it took only a short time for wood cutters to destroy every vestige of timber close to town; a dilemma recognized by all. "There is but very little wood of any sort available for coal to be met with on the hills within a radius of 25 miles of Eureka," commented the *Eureka Sentinel* in spring of 1874, "and the area of the circle is every day rapidly extending itself."

As the *Mining and Scientific Press* noted, depletion of fuel was not just a Eureka problem:

> A difficult problem for Nevada State in the near future is that of the fuel supply for its immense mining enterprises. Its limited good forests along its western line, on the eastern narrow slope of the Sierras, is being rapidly swept away, while through its central portion a stranger would think there is none, so little is there to be seen. Virginia City with its immediate vicinity consumes in round numbers some 300,000 cords a year, or at 30 cords per acre, 10,000 acres. Thus in a few years would lay bare all that portion of the Sierra range in Nevada State. But Virginia City is only a part of the mining region of Nevada. The Eureka and Richmond mills alone, besides all the different hoisting works at the six or seven mines, use up daily and in round numbers 180 cords, i.e., together 360 cords. This is mostly in the form of charcoal hauled in by teams from many miles away for smelting all their ore. This is about 130,000 cords a year. Tybo has used three-eighths as much, or 50,000 cords.
>
> This will, not long in the future, leave the State without a forest in its territory, except the very slowly-maturing second growth. Already sage brush is being extensively used, but this, too, must fail in time. At Lewis, its two mills use mostly this brush for fuel at an expense (for one of the mills) . . . of $34 a day by contract for all they might need, though some five cords of wood were also consumed. At Tuscarora also sage brush was mostly used, though its extensive

> valley in front is all cleared of this brush for nearly 10 miles away. The second growth come forward very slowly, taking probably 20 years for maturing. Still, in Tuscarora a large amount of wood has been burned. The Grand Prize even now, at its new hoisting works, alone uses 30 cords of wood a day, at an expense of $17 per cord on the average. As yet no good coal has been found west of Salt Lake and east of the Sierras. . ."[24]

This was a situation the whims of nature could and did upset.

The winter of 1873-74 was especially severe, retarding wood harvest and charcoal burning. By spring only 175,000 bushels were ready. Hardly enough fuel to sustain operations for the next eight months considering that Eureka's eight furnaces would consume 3,264,000 bushels during that period. Even with introduction of several conical kilns by Reilly & Lockwood, the chief supplier of charcoal at Eureka, the situation could not be corrected.[25]

Building of the Eureka & Palisade Railroad momentarily eased the fuel crisis. Connecting with the Central Pacific at Palisade, 90 miles north of Eureka, this narrow gauge line was extended to Alpha by end of 1874, allowing a limited supply of coke to be brought in from the Pancake range near White Pine. Teamsters picked the commodity up at Alpha, the midway point, and laid it down at Eureka.

Although the Eureka & Palisade Railroad reached Eureka on October 22, 1875, not much coke was brought in, and the smelters went back to the old standby of charcoal, purchasing it from local charcoal burners, and making up the shortfall from suppliers in Utah and Wyoming. Nevertheless, the price of charcoal crept upward until it stood between 27 and 28 cents a bushel, a price smelter operators considered exorbitant.[26]

It all came to a head during the "dull times" of 1879, when mines and smelters throughout the West were closing due to declining metal prices. That summer Eureka charcoal burners, who had long suffered at the hands of dishonest teamsters and merchants, organized to alleviate their poverty. On July 6, 1879, 500 charcoal burners met in Celso Tolli's saloon in Eureka to formulate some sort of organization. Guido Bassetti was elected temporary chairman and a committee appointed to negotiate a higher price for charcoal with the Eureka Consolidated and Richmond companies.

Before the superintendents of both companies, who were present at Tolli's saloon, the burners claimed they could not make a living at the present per bushel rate of 28 cents. When they requested the price of charcoal be raised to 30 cents a bushel both superintendents balked at the suggestion. Edward Probert of the Richmond Company said his company's obligations to stockholders made a move upward in price impossible, and Henry Donnelly of the Eureka Consolidated stated he had already contracted for his charcoal and that he would close his furnaces unless he got the fuel at the contract price. Both men stated any increase in the price of charcoal would result in smelting loss.

Rebuffed, the charcoal burners adjourned until the next day. On the evening of July 7, 1879, charcoal burners assembled at the Eureka Opera House, the only place in town large enough to seat a large congregation. The meeting started off with election of permanent officers: Joe Maginni, president; Lambert Molinelli, vice-president; Joe Hansman, secretary; and Sol Ashum, treasurer. A name was chosen for their organization — the "Eureka Coalburners Protective Association." And a committee was appointed to induce all burners of the area to join the organization. Then the coalburners got down to business.

The burners resolved to raise the price of charcoal to 30 cents a bushel. If their demand was not met they threatened to cut off the supply of fuel. There was more to their demand, however. Not only did they want an increase of a few cents in the price of their commodity, they sought to correct a number of injustices. They complained there was collusion between smelter superintendents and teamsters to cheat the charcoal burners, a charge with foundation.

Although in February 1879 the State Legislature had fixed a standard bushel of charcoal at 2,747 cubic inches and stipulated that all fuel bought and sold must be so measured, the measurement was manipulated. Teamsters who delivered the charcoal to the furnaces cheated burners out of revenue by not showing smelter receipts. Then too, the smelters did not deal directly with the burners, but forced them to take their pay in orders for goods on certain merchants who, when the orders were presented, charged higher prices than were asked of cash customers. To rectify this, the burners resolved in the future to take nothing but cash for their labor. On July 8 the Eureka Coalburners Association drew up the following notice and published it in the *Eureka Daily Sentinel* the next day; it ran for ten days:

> Take notice, that on and after this date, no coal will be sold to anyone by the members of the Eureka Coal Burners Association unless the same is disposed of at the furnaces for the sum of Thirty Cents per bushel.

The notice and the actions of the Association caused mixed reactions. Edward Probert, manager of the Richmond furnaces, considered them "bulldozing" tactics, and threatened to close down the smelter. The Eureka Consolidated was more conciliatory, offering instead to pay 27½ cents per bushel. Because both companies had stockpiled about 120,000 bushels of charcoal each, Eureka's business community assumed the burners could not last long without income, even if they managed to shut off the supply.

The Coalburners' Association, however, was determined to force the issue. By July 16 its membership stood at 1,196, and organizers were at work recruiting the rest of the estimated 2,000 men engaged in the coaling profession in the Eureka District. The next day groups of burners, many of them armed, began barring passage of teamsters to charcoal burning grounds. Three days later they picked on teamster Robert Brown. When five armed Italians prevented him loading charcoal near the Willows, Brown returned to Eureka and filed charges of malicious mischief against the Italians. Hopping mad, Brown filed suit on July 24 against officers of the Association, claiming four men, as officers of the Association, "had passed, or caused to be pass, resolutions interfering with and preventing him from fulfilling a contract with Reinhold Sadler." Brown sought $3,850 in damages. This suit apparently caused dissension within the ranks of the burners, for Maggini and Hansmann — "intelligent, well-meaning men who had the interests of their countrymen at heart" — resigned as officers of the Association on August 5.[27]

That day the Coalburners' Association elected new officers: Angelo Dilboni, president; Guiseppi Martinoli, vice-president; Guido Bassetti, secretary; Niceleo Ratti, assistant secretary; and Severino Strozi, treasurer. More militant than the original officers, these men overruled members who favored 28 cents a bushel, a price that smelters were willing to accept. To their way of thinking negotiation with the mining companies was out of the question. Association hardliners had decided no burner would be allowed to deliver at a price less than 30 cents and that all burners had the privilege of seeing furnace receipts.

The 30 cent a bushel price of charcoal, however, was not the real issue. What militant members wanted was abolition of all teamster contracts and the right to deal directly with the smelters. To attain those goals the Association resorted to a "shotgun policy" of intimidation. During the following week bands of armed Italians, with "the avowed intention of preventing any and all teams from loading charcoal until furnaces came to their terms and paid the rate," took possession of every loading point in the district. They interferred with loading of charcoal, destroyed property and threatened lives of peaceful citizens. At Joseph Tognini's ranch near Alpha, 30 miles northwest of Eureka, burners scattered 2,000 bushels of charcoal through the sagebrush. They also took possession of an Alpha saloon but retreated to the hills on rumor of an approaching armed force.

Operating in small bands, communicating through a system of spies and runners, burners prevented the loading of charcoal throughout the district. Intimidation progressed to thuggery on the morning of August 9 when six coal burners entered the home of teamster George Lamoureux, dragged him from his bed and threatened to beat him if he persisted in hauling charcoal. George, however, was a man who did as he pleased. Later that day he drove to Cassini's ranch to pick up a load of charcoal. There he was confronted by armed Italians. Although armed himself, he returned to Eureka with empty wagons. Just as Brown had done, Lamoureux went to the Sheriff's office and swore out warrants against those who prevented him from loading. Deputy Sheriff Simpson immediately left town to serve the papers, but was unable to locate the culprits.

Up to this time many Eurekans supported the coal burners' cause, but with coercion, violence, and destruction of property, toleration vanished. The *Leader* suggested that a large posse be sent out with the wagons to each area where a large amount of charcoal was ready for loading. The posse could protect the teamsters until all fuel was removed, then transfer operations to another coaling point. On August 11 the Eureka Consolidated announced it would not be dictated to, and threatened to shut down the furnaces if it did not get a supply of charcoal at the desired price.

All this infuriated Sheriff Matt Kyle. In a council of war on the afternoon of August 10 he and his deputies plotted a course of action. His solution was to raise a posse of 100 well-armed men, a number deemed sufficient to cower the Italians. That plan fizzled when only 35 men responded, forcing Sheriff Kyle and B. J. Turner, chairman of the Eureka Board of County Commissioners, to telegraph the following message to Governor John H. Kinkead at Carson City on the morning of August 11:

> 2,000 persons, banded together, and with arms in their possession, defied the civil authorities, and refused to have any of their number arrested. They now hold forcible possession of many coal pits

> in this county. By force they have prevented, and are now preventing the owners of charcoal from hauling it to the furnaces, and they threaten to destroy other property and burn the town. Arrests have been resisted by the rioters who are well armed and organized under the command of desperate leaders.

Governor Kinkead wired Eureka for more information on the state of affairs and the availability of arms and ammunition in the district. Colonel Edward N. Robinson of the militia answered, stating there were 1,000 men at Alpha, Roberts Creek, Fish Creek and Newark who would resist authority, and at least that many more at various points who would assist in case of trouble. The Governor was urged to take immediate action as there were only 100 rifles available and little ammunition. Robinson informed the Governor that members of the Second Brigade of the State Militia were standing ready at the armory. The Governor approved the activation of the militia and sent arms and ammunition on the evening express. By evening of August 11 the full brigade was assembled at Eureka's armory.[28]

Activation of the militia lessened the bravado of the Italians, and lawmen apprehended a number of their leaders without resistance. On August 13, Angelo Dilboni, president of the Association was arrested. A justice of the peace charged him with fermenting disturbances and inciting riots, set bail at $1,000, and scheduled his trial for September. Severento Strozi, treasurer of the Association was also arrested and lodged in the county jail. More arrests were made at Fish Creek and Alpha on August 14, and the next day the jail was so full of Italians that some were placed in the armory.

Although a few Association leaders were in jail, groups of burners still disrupted the loading of charcoal to the extent that many Eurekans felt they should be taught a lesson; no tears shed if a few Italians were shot. The two mining companies added to the tenseness of the situation by announcing they were shutting down their furnaces. On top of that, General George M. Sabine proclaimed the militia would be used only if civil authorities lost control of the situation. In the mind of Sheriff Matt Kyle that was not going to happen. He would do all in his power to keep fuel flowing to the smelters, and to protect teamsters and their property.

On the morning of August 18 Sheriff Kyle assembled his posse: Deputy Sheriff Simpson, special deputies Joseph Toomey, William Martin, Robert Brown, Hank Storey, Jim Porter, and G. H. Smith. Also present and deputized was Marshall Rich, foreman for teamster

George Lamoureaux. Handing over John Doe warrants to Deputy Simpson for the arrest of leaders of the coal burners, Kyle instructed his deputies to protect the teams of George Lamoureaux.

At dusk on August 18 the posse arrived at the coaling ground about seven miles south of Fenstermaker's ranch on Fish Creek, where they were confronted by about 100 charcoal burners, some of whom were armed.[29] There are two versions of what happened next. One version is that when the Italians forbade the loading of charcoal, Marshall Rich who had brought along several wagons, told them "he had come there for that purpose; that they had no right to the coal; that the men who burned the coal and owned it wanted it loaded, and that he proposed to see that it was loaded." He then told the owners of the coal to commence loading, which they attempted to do but were pushed back by the leaders of the gang, and commanded not to touch a sack. Angry, G. H. Smith and Rich then began to load the sacks upon the wagons. They were ordered to stop and pushed aside by the Italians. When they again tried to load, the leader drew a pistol and fired at Rich, the bullet grazing the deputy's head. At the same instant, another striker rushed upon Smith with a long knife, which caused lawmen to fire with their revolvers.

The other version is that lawmen demanded coal burners produce some of their leaders to be taken back to Eureka for trial. When the ultimatum was refused lawmen opened fire. The burners scattered in all directions leaving five of their comrades dead and six others wounded.[29] Expecting the Italians to continue the fight, the officers did not pursue the retreating strikers, but stayed by the wagons. Fourteen prisoners were taken but only three were brought to town, as the rest were found to be unarmed.

The battle not being renewed, the posse left for town. Sheriff Kyle, Deputy Storey, Dr. Thomas and Coroner Smith returned to Fish Creek about three o'clock the morning of August 20. They examined the dead, but found none of the wounded men. An inquest was held and a Coroner's Jury brought in a verdict that, "the deceased persons came to their death from shots fired by the Sheriff's posse while in the discharge of their duty."[30] On September 5, 1879, a Grand Jury expressed "their entire approval of the action of the Sheriff's posse in their determination to enforce the law at all hazards." Thus the Eureka Coalburners' Protective Association was squelched, the Fish Creek War, as it was called, ended, and life at Eureka returned to normal. "A noticeable and pleasant fact is the revival of trade in Eureka," the *Leader* noted, "save the unfortunate coal troubles, there is a feeling of confidence in every direction, based upon the apparent stability and improved prospects of the mining interests."[31] The *Eureka Sentinel* estimated the cost to the county of the coal burners war at from $10,000 to $30,000."[32]

* * *

If Western smelting was to remain profitable something had to be done to replace the dwindling supply of charcoal. The solution was importation of another fuel, and in the 1870s that was coke, available only from Connellsville, Pennsylvania, at $34 a ton. Nevertheless merchants and smeltermen sought the commodity. In May 1872 Salt Lake merchants Gordon and Murray received five carloads of coke — the first in Utah — which they sold to local smelters. The following year they brought in another 5,000 tons. Amounting to about $170,000 worth of fuel, this tonnage in coke wetted the appetites of railroads, and by end of 1873 there was considerable competition between carriers to furnish Great Basin smelters with Pennsylvania coke. Meanwhile, a search was launched in Utah for coking grade coal.[33]

Coal existed in Utah, and was used by Mormons as early as 1854. But it was assumed, "upon very imperfect experiment," that Rocky Mountain coal would not coke. Some men doubted that, among them Amos Stowe, who had come to Salt Lake City from San Francisco to enter the mercantile business. In April 1872 he leased an outcrop of bituminous coal (the Hinton Mine) near Evanston, Wyoming, employed some coke burners and began experimenting with conversion of coal to coke upon a large scale. Successful, Stowe proved that Rocky Mountain coal could be converted to coke.[34]

The following year (in late 1873) coke was made from lignite found at Trinidad, Colorado, about 90 miles south of Denver — the first good coke for smelting purposes ever made from lignite alone in America." Prior to that it had always been necessary to mix bituminous coal, tar, or similar materials with lignites, to produce a coke, which in most cases was an indifferent fuel for shaft furnaces. In November 1873 Frederic A. Eilers pronounced Trinidad coke the "answer for all purposes of lead and copper smelting in shaft furnaces, and if made in proper coke ovens will probably be sufficiently dense to carry the high smelting column of the iron blast furnace."[35]

That endorsement caused the Denver and Rio Grande Railroad

to push its line to Trinidad six months later. Aware of the rise in the price of charcoal, J. C. Cameron, Jr., inventor of the beehive kiln, stated in July of 1874 that "lignite and coke would have to replace charcoal if smelting was to be economically carried forward in Utah." He viewed the coal measures of Sanpete Valley in Carbon County, 40 miles south of Provo (then the terminus of the Utah Southern Railroad), as a source for "cheap and efficient fuel." Geologists A. P. Bouton and J. E. Clayton found Sanpete Valley coal "highly inflammable and richly bituminous in character, admirably adapted for reverberatory and blast furnace purposes."[36]

In May 1874 the Fairview Coal Mining and Coke Company was organized by Perry Crabbs (president), D. H. Bentley (vice president), William B. Welles (secretary and treasurer), and H. Carlston (superintendent). Capitalized at $1,000,000, the company opened seven mines, and erected coking ovens in the upper reaches of Huntington Canyon. A small community was founded, called Connellsville after the Pennsylvania coke-producing locality. A lot of coal was mined and for a while coke was converted at a rate of 12½ tons a day.

Practical working tests conducted at the Germania Refinery found Fairview coke "equal in all respects to the imported article from the East." There were a number of problems, however. Although Fairview coke measured up to the Pennsylvania product, there was not a lot of coal of coking grade in the mines. It was lack of a railroad link from the mines to Springville on the Utah Southern line that doomed the company. Nevertheless, this was the first commercial use of Utah coke in Salt Lake Valley smelters. The Fairview experiment stimulated commercial interest, and for a while it was thought a large supply of coke could be placed at reduction works at a cost not to exceed $20 a ton.[37]

The Charcoal trade lingered until the mid-1880s when it was severely impacted when coke was introduced into Utah from Crested Butte, Gunnison County, Colorado. A few years later, exploration by the Pleasant Valley Coal Company revealed deposits of coking grade coal near Castle Gate, within easy access of the Denver and Rio Grande Western Railroad, which had taken over the Utah Southern Railroad. By 1896 three mines had been opened and 104 coking ovens built; those ovens sounded the death knell of Utah's charcoal trade.[38]

* * *

1. *Tribune*, April 26, 1871; May 24, 1872.
2. Reprinted in *Mining and Scientific Press*, May 9, 1874.
3. *Ibid.*, May 9, 1874.
4. *Ibid.*, November 3, 1877.
5. *Engineering and Mining Journal*, August 15, 1871, p. 105.
6. James B. McNair, *Simon Cameron's Adventure in Iron, 1837-1846*. Los Angeles: by author, 1949.
7. Eli Bowen, *The Pictorial Sketch-Book of Pennsylvania*. Philadelphia: 1852, pp. 117-18.
8. T. Engleston, "The Manufacture of Charcoal in Kilns," Part One, *Mining and Scientific Press*, November 3, 1883; originally published in *Transactions*, American Institute of Mining Engineers, Vol. VIII, 1880.
9. Although the principle of all conical kilns were the same, construction varied in different localities. At Wassaic, New York, walls were 30 inches thick at the bottom, diminishing gradually to 14 inches at the top, which was made flat for a cast-iron vent-hole. At Norton's Iron Works, near Plattsburg, New York, the wall were built with a batter three inches to the foot up to five feet. At this point, the height of the kiln being determined as 20 feet, a perpendicular is raised, and somewhere on this line a center is found from which an arc of a circle will meet the flange of the charging hole at the top, which is made of a cast-iron ring four feet in diameter and eight inches deep, projecting six inches over the top. This makes the wall a little thinner at the top than the bottom. The flange of the ring is normal to the curve of the masonry, which is generally built of red brick.
10. *Tribune*, August 9, 1874.
11. *Ibid.*, January 14, 1873.
12. The Sultana Works needed only 600 bushels of charcoal per day.
13. *Tribune*, November 19, 1872; August 9, 1874.
14. *Ibid.*, July 25, 1874.
15. *Ibid.*, January 14, 1873.
16. The Hilliard flume was twice as long as the flume which conveyed timber from the crest of the Sierra Nevada to lumber yards at Carson City. That flume was 12 miles in length, V-shaped, with a carrying capacity of 600 cords of wood a day. *Mining and Scientific Press*, March 16, 1895. By 1878 the Hilliard flume had been dwarfed by others in California. The Butte Lumber Company, east of Chico, operated a 35 mile long flume, and there was a flume at Red Bluff 32 miles long. The Sierra Flume & Lumber Company had 150 miles of flumes. *Scientific American*, Supplement No. 116, March 23, 1878, p. 1840.
17. In the summer 25 men were employed on the flume, 25 at the sawmill and an average of 125 in the timber.
18. *Utah Mining Gazette*, February 14, 1874; *Tribune*, January 1, 1877; January 1, 1883.
19. *Tribune*, January 1, 1878.
20. *Mining and Scientific Press*, July 15, 1876.
21. In spring of 1877 ten conical kilns were built at the head of Wildrose Canyon in the Panamint Range of Inyo County, California, to fuel the two smelters of the Modock Consolidated Mining Company, owned by George Hearst. Twenty-five feet high and 30 feet in diameter, these kilns held about 42 cords of wood each. Under supervision of saloon keeper James Honan, they produced charcoal during spring and summer of 1877. Depletion of the luxuriant growth of nut pine, cedar and juniper put an end to Wildrose Canyon charcoal business. *Mining and Scientific Press*, September 22, 1877; Richard E. Lingenfelter, *Death Valley and the Amargosa. A Land of Illusion*, pp. 26-27.
22. Emmons and Becke, *Statistics and Technology of the Precious Metals*, p. 471. *Tribune*, July 24, 1878.
23. See railroad tonnage charts in *Tribune*, January 1, 1880; January, 1, 1887.
24. Issue of August 23, 1879.
25. Reprinted in *Mining and Scientific Press*, May 9, 1874.
26. *Ibid.*, May 9, 1874.
27. *Eureka Sentinel*, August 6, 1879.
28. *Tribune*, May 24, August 12, 14, 19, 1879.
29. For a thorough discussion of the charcoal burners war at Eureka consult: Phillip I.

Earl, "Nevada's Italian War," *Nevada Historical Society Quarterly*, Vol. XII, No. 2, Summer 1969, pp. 47-87. In the same issue of the *Quarterly* Wilbur Shepperson's "Immigrant Themes in Nevada Newspapers" adds additional details.

30. The dead were: Pompeo Pattini, 35; Giornici, Switzerland; Antonio Canonica, 25; Corte Ciasea, Switzerland; Marcellino Locatelli, 25; Mozzio, Italy; Theodoro Zerli, 28; San Fellipo, Italy; and Giovanni Pedroni, 23, Italy.

31. Angel, *History of Nevada*, p. 438.

32. *San Francisco Chronicle*, August 20, 1879; Angel, *History of Nevada*, p. 438.

33. *Tribune*, August 28, 29, 1879.

34. *Utah Mining Gazette*, December 6, 1873.

35. *Mining and Scientific Press*, April 20, 1872.

36. *Utah Mining Gazette*, November 29, 1873; July 18, 1874. Bouton and Clayton submitted San Pete coal to chemists in the United States and England for testing. They found the coal had superior coking qualities, viz: coke, 62%; volatile matter 28%; ash 10% = 100%.

37. *Utah Mining Gazette*, July 18, 1874; J. H. Morton, "The Coal Mines of Central Utah," *Engineering and Mining Journal*, February 3, 1877, pp. 76-77; August 8, 1914, pp. 248-253.

38. *Tribune*, January, 1887; Ronald G. Watt, *A History of Carbon County*. Salt Lake City: Utah State Historical Society, 1997, p. 107-11.

Chapter Six

THE ERA OF THE PORTABLE FURNACE, 1880-90

As related in previous chapters, John Williams, Jr., made his mark in Western metallurgy first at White Pine and Eureka, Nevada. He then showed up at Stockton, Utah, in spring of 1873 to put the Chicago Company on a firm footing. Accomplishing that he pushed on to Tintic and Milford, Utah. John prospered in Beaver County for a few years running his smelter and a hotel at Milford. The building of the Francklyn Smelter by the Horn Silver Mining Company put Williams out of business in Utah. It is doubtful he shed any tears over that, for shortly thereafter he went to San Francisco, where his father had teamed up with DeWitt F. Bisbee, his brother-in-law, to fund and promote mining and other commercial ventures.

Bisbee and Williams was a good business combination. The senior Williams had the mining and metallurgical expertise, and Bisbee was a well-known contractor who made a fortune building seawalls and filling in San Francisco's marshlands. DeWitt was also associated with Martin, Ballard & Company, a forwarding and commission house which in the early 1870s was agent for Rock Springs and Evanston coal, Connellsville coke, and Wyoming hematite, essentials for the smelting industry of the Great Basin.

John Williams, Jr., brought to the firm of Bisbee and Williams a decade of metallurgical experience, hard-won in the base metal districts of Nevada and Utah. He had built every conceivable type of smelter and experimented with the chemistry of slags; he understood the economics of metal reduction. By 1878 John was aware that a drop in metal prices was taking the profit out of argentiferous lead production. Gold was displacing silver in miners' preference, and copper was on the rise due to the spanning of a continent with telegraph lines and the application of the red metal to building trades. While he may not have

The Rankin-Brayton smelter represented an improved style of water-jacket smelter. Introduced in 1881 it embraced features not found in other furnaces, the most notable being great economy and capacity for continuous work. With ordinary care, this smelter could run for six to twelve months with little or no interruptions. This smelter was shipped complete, ready to set up; an obvious advantage over furnaces that required brick or stone work. Another advantage was its distribution of blast, which avoided dead centers, thus preventing the tendency to freeze. Near perfect combustion assured great fuel economy, and produced a clean, high grade bullion or matte. (*Mining and Scientific Press,* June 18, 1881)

been clairvoyant enough to foresee the Electric Age, his sixth sense told him what western mines needed most was a cheap, portable water-jacket furnace, which with slight modification could process small, rich deposits of lead or copper carbonates and oxides.

Sometime in 1878 or early '79 John collaborated in designing a 30-ton capacity furnace with older brother Louis, who with brother Ben had smelted copper at a locality 60 miles south of St. George, Utah. (Father may have also had a hand in the project.) What they produced was a simple piece of equipment. The Williams brothers merely elongated the German cupola furnace, placing around the crucible an iron envelope. Between that and the furnace wall water circulated. Addition of a water-jacket, together with cast-iron boshes, made the furnace virtually indestructible. To build a prototype furnace the Williams brothers enlisted the aid of Ira P. Rankin and A. P. Brayton, owners of the Pacific Iron Works; men who had years of experience fabricating all varieties of mineral processing equipment.

The Pacific Iron Works was a pioneering company. Founded in 1850 by Robert Egery and David B. Hinckley, it was first located at 127 and 131 First Street, between Mission and Howard, in what would become San Francisco's foundry district. Three years later the interests of the founding partners were purchased by New Englander E. B. Goddard, who enlarged the company's manufacturing capacity. The Pacific Foundry and Machine Shop started out doing general foundry work and maritime repair. That changed in 1862 when Ira P. Rankin, a Scot boilermaker, and A. P. Brayton, a molder from Rhode Island, bought into the business. Two years later, on April 25, 1864, the company was incorporated under the name of Pacific Iron Works, with a capitalization of $200,000, divided in 1,000 shares at $200 each.

Focusing on repair of marine engines and ship hardware, the firm built up an impressive clientele which included the Ben Holliday Steamship Line and the Pacific Mail Steamship Line. This repair work blossomed into full scale fitting out of ships, and by 1865 Pacific employed 150 men who kept three cupolas going, melting about 20 tons of iron at any one time. That year the company built machinery for the Steamer *Del Norte*, and overhauled the *J. T. Wright* which cruised the north Pacific in service of the Russian Telegraph Company.[1]

With opening of trade with China and Japan and the increasing coastal business, which created a demand for new vessels, the firm branched into design and construction of ocean-going vessels. Concurrent with its maritime business, the Pacific Iron Works performed cus-

The Rankin-Brayton smelting plant in full running order; the blast furnished by a Baker blower. Powered by an independent small engine, the crusher was of sufficient capacity to crush in a few hours all ore and flux required for the entire day. A horizontal, tabular boiler furnished steam for engines and pumps. The whole structure was made complete at the shop — mostly of wrought iron, to insure strength and lightness, and in sections to facilitate transportation and set-up. No brickwork was required except for crucible and two courses on top of jacket, making the cost of erection less than half that of any other style of smelters. A boiler-iron foundation and curb was provided, atop of which the whole structure was built.

tom work for mining companies, building engines for mines on the Comstock lode. In 1860 Pacific introduced the Bryan Quartz Mill, and followed shortly thereafter with the Varney amalgamation pan. Both appliances were widely accepted, particularly in Kern County, California, and in Nevada. In spring of 1867 Pacific made a novel departure in the engineering of quartz mills, building under the patent of James and Nathan Condict, Jr., a mill where steam was applied directly to the stamps.[2]

The Pacific steam mill consisted of a battery of two stamps, the downward movement of which was effected by aid of high-pressure steam, while the upward movement was produced by steam at low pressure. The stamp heads were affixed directly to the pistons. Induction and eduction of steam to the cylinder's upper portion, which produced the "strike" of the stamp, was made through a direct communication with the boiler, and could be kept at any desirable pressure. Lightweight, only 1,500 pounds without boiler, the machine was intended as a prospecting mill. Including stamps, mortar, and cylinder, it was a mere two feet square by six feet high.[3] A few years later, Pacific redesigned this mill, and marketed it as the Wilson Steam Stamp Mill, three of which were constructed in 1870 for the Overman Mine at Virginia City, Nevada.

By 1867 construction of mills and Varney pans constituted a large portion of Pacific's mining business; so much so that three years later Pacific's foundry was entirely rebuilt. A three story brick building, "perhaps the best in San Francisco at the time," was erected to house a pattern-making shop, foundry, and machine shop, the latter completely refitted with new tools. The revamping enabled the firm to turn out the heaviest equipment.

In the early 1870s the company excelled in the design and building of mills and hoists, much of the equipment tailored for the environment in which it would be set up. Stamp mills and hoists were designed to be sectional, no piece weighing over 200 pounds, allowing for mule transport into rugged terrain. Little wonder, therefore, that mining companies in Mexico and more isolated regions of Western North America sought Pacific's machinery.

While Pacific was perfecting hoists and milling equipment, it developed a line of furnaces. Working with John Howell at the Comanche and Indian Queen mines at Benton, Mono County, California, the foundry fabricated a rotary furnace for chloridizing silver ores and sulphurizing copper, gold, lead, and zinc ores. The Howell Rotary Furnace, as it was called, was no more than a revolving tube mounted on carrying wheels, with one end opening into an upper furnace chamber, the other into a lower furnace into which ore was discharged from the tube after treatment. The tube was made in sections having flanges, the upper section being smaller than the lower one, and connected by a mid-section of intermediate size.

As early as fall of 1870 Pacific was fabricating slag pots and shipping furnace components to Eureka, Nevada.[4] In June 1871 the *Mining and Scientific Press* reported the Pacific Works "turning out considerable iron work for smelting furnaces at Eureka, White Pine, and Owens River."[5] It also experimented with copper-lined furnaces, cooling them by applying an iron-jacket around the tuyere section and forcing air into the intervening space. The company must have had some success with these air-cooled furnaces, for in 1878 it marketed them to the Eureka Consolidated Company at Eureka, Nevada, and to W. H. Raymond of the Lodi District, also in Nevada. Another furnace was installed at the Modoc Mining Company of Inyo, California.

Pacific did not hit upon a commercially successful smelter until Ira B. Rankin put a steel jacket around Beaudry's furnace at Cerro Gordo in 1874. Little wonder, therefore, that John and Louis Williams and the Pacific Iron Works were a perfect match; their product, called the Pacific Furnace or the Rankin-Brayton Furnace, was an instant success.

The Rankin-Brayton Furnace embraced many features not found in other furnaces, the most notable being great economy and capacity for continuous work. With ordinary care, the smelter could run for six to 12 months with little or no interruptions. This smelter was shipped with all components — boiler, steam engine, pumps, fan or blower, rock breaker, tools, assay kit, and piping. Ready to set up, it had an obvious advantage over furnaces that required brick or stone work. Another advantage was its distribution of blast, which avoided dead centers, thus preventing freeze-ups. Near perfect combustion assured great fuel economy, and produced a clean, high grade bullion or matte.

The Rankin-Brayton Furnace was tested first on copper ore at the San Francisco Copper Company of Nevada County, California, in January 1880, and then tested in Nevada on lead ore shortly thereafter.[6] Arizona Territory, however, would be the proving ground for the furnace, for by fall of 1880 Pacific had installed a number of furnaces in the territory along the line of the Southern Pacific Railroad: two at the Castle Dome Smelting Works above Yuma on the Colorado River; one at the San Xavier Copper Company south of Tucson; and another for a New York company operating near Gila Bend.

How efficient these furnaces were is attested to in a letter to the manufacturers by W. P. Miller, superintendent of the Castle Dome Smelting Works (dated February 18, 1882):

> . . . we have two of your 40-ton galena smelters in our works that have been running most of the time for the past two years. With ordinary care I have no difficulty in keeping them in condition for constant service. The patent circulating plates, recently introduced into your water jackets, I regard as the most important improvement ever made in water jacket furnaces, and worth more than the entire cost of an ordinary jacket.
>
> There is no question but that a jacket, for either galena or copper ores, with this improvement, will last two or three times

> as long as without it, besides lessening the liability to injury and the consequent expense and loss of time for repairs. Since the adoption of this improvement, I have had no trouble whatever with my jackets. The circulation of the water is so rapid that the inner surfaces of the plates are kept scoured and entirely free from any incrustation, and it seems to afford perfect protection from the effects of mineralized water, and from the overheating and burning out so common in all other jackets. Your style of sectional water jackets have many and obvious advantages over all other forms. The distribution of the blast and the angle of the tuyeres are such as are calculated to secure perfect combustion and the greatest economy of fuel.
>
> With a long and varied experience in smelting ores from a wide range of mineral country, I unhesitatingly give your furnaces the preference over any I have ever seen.[7]

The most successful application of the Rankin-Brayton Furnace was the one installed by the Williams brothers in the Mule Mountains of southeastern Arizona. That story began on May 12, 1880, when John Williams, senior member of Williams and Bisbee, advised Martin and Ballard to bond for $20,000 the Copper Queen and Copper King claims, two properties owned by Edward Reilly. A deal was struck whereby Williams, Bisbee, Martin, and Ballard would furnish capital to work both claims and to install a first class smelting works in return for seven-tenths interest in the Copper Queen Mine and refinery, and two-thirds of the Copper King Mine. Reilly would hold the remaining interest in the properties.

A 36-inch furnace was shipped to the Mule Mountains. Costing $11,000, it was the largest manufactured at the time by the Pacific Iron Works, having a capacity of 30 tons depending on fusibility of charge, strength of blast, and water supply. It was of unit-design, required no brickwork, and arrived in the Mule Mountains ready to set up and operate as soon as plumbed to a water source. Its auxiliary equipment consisted of a No. 4½ Baker blower and a Blake crusher. Power to run these units came from separate steam engines (a 10-horsepower engine for the breaker and a 12-horsepower engine for the blower). A tubular boiler and steam pump completed the plant. Although the furnace required 25,000 gallons of water every 24 hours as coolant, addition of a reservoir and cooling tank cut water consumption to 3,000 gallons a day.

In less than 60 days from finalization of contracts the Rankin-Brayton Furnace had been shipped by rail from San Francisco, unloaded at Pantano station, the Southern Pacific terminus at the time, and transported by Jimmy Carr of the Cerro Gordo Freighting Company 60 miles along the San Pedro Valley, hauled around the southern end of the Mule Mountains and up Mule Canyon to the Copper Queen Mine. The furnace was installed directly below the open cut which constituted the mine, so that ore could be shot directly through the breaker shed to the smelter.[8]

While 46-year-old Louis Williams supervised erection of the furnace, Reilly, Martin and Ballard began enlarging the mine. For a month nearly 100 men labored at the Copper Queen Mine. A cut 50 feet wide was started across the strike of the vein, and extended 100 feet into the hillside. Penetrating a cap of limestone, the miners exposed so solid a body of ore that it was literally "quarried out by the ton." By August 12, according to an *Arizona Daily Star* correspondent, "1,300 tons of ore had been taken out which will work twenty-five percent, and an estimated $200,000 worth or ore [was] in sight." All that was needed now was to fire up the smelter, and that was not long in coming.

Already 10,000 bushels of charcoal was at hand and a crew of Mexicans was burning 25,000 bushels more. Five carloads of English coke had been brought from San Francisco to be mixed with the charcoal. The smelter was finished by mid-August, plumbed to Malvina Spring near Castle Rock, and brought to operating temperature. Insufficient flow of water from spring to tank caused the first furnace run, on August 20, 1880, to fail. Installation of larger pipes remedied the situation, and a month later the smelter was producing a pig of black copper, weighing about 180 pounds, every 15 minutes. The result of the first 14 days of constant operation was more than 55 tons of copper, 90 percent pure, which was loaded aboard Remi Nadeau's 20-mule team freight wagons and shipped to Pantano station.

The smelter added impetus to other mining ventures in Mule Canyon. The Neptune Mining Company, consisting of seven claims southeast of the Copper Queen, was enlarged. Two shafts and an adit was dug to tap a rich body of ore. The Excelsior, an adjoining claim belonging to the same company, had an open cut 18 feet wide, "every foot in good ore."

Capital for working the Neptune venture was subscribed in Hartford, Connecticut, by attorney William Herring who had come west from New York. Because the Copper Queen smelter monopolized the only source of water in Mule Canyon with enough volume to cool a refinery, the Neptune Company had to seek a smelter site 15 miles away on the San Pedro River. Six hundred acres of the San Rafael Grant was purchased, and on the edge of a mesa 500 yards from the east bank of the river, August Raht, who later earned the title of "prince of lead smelters,"[9] erected a 30-ton water-jacket furnace. The smelter gave birth to the small settlement of Hereford. Company offices and a large adobe dwelling for the Herring family, known as the "Castle," were erected at the Neptune Mine in Mule Canyon.

Although the Herring property showed rich ore near the surface, the lense of copper carbonate pinched out. With capital depleted on the smelter and the "Castle," exploration for other orebodies could not be carried out and operations ceased. The property went into receivership and was sold at sheriff's auction in 1886 to the Holbrook and Cave Company. Ten years later it was incorporated into the general holdings of the Copper Queen Mining Company.

In the meantime, luck was on the side of Reilly and his associates. On the northern slope of Queen Hill, the Copper Queen Company trenched, drilled, and blasted its way into an almost circular body of ore, 60 feet in diameter, which dipped into the hill at a 30 degree incline. This ore mass was unbelievably pure and in places breathtakingly beautiful. Crystallized masses of malachite looked like green velvet, and vugs lined with sparkling prisms of deep blue azurite, were encountered in the brown ledge-matter. More often the copper carbonate appeared as botryoidal folds and chunks associated with ruby-red cuprite, a copper oxide.

The gigantic tube of ore, enclosed by marbleized limestone, was so basic that quartzite, gathered in an adjacent canyon, had to be added as flux to the furnace charge. For months the smelter devoured thousands of tons of ore that ran 23 percent copper; and with a single fusion a black matte was produced which ran 94 to 96 percent metal.

At first the deposit was followed by a single drift. In December 1880, however, the Copper Queen Company commenced another cut 15 feet below the previous one which revealed, according to the *Arizona Daily Star*, "some of the richest and most beautiful specimens ever taken out of the mine." By January 1881 the drift had been extended 200 feet into the mountain, and on the left side of the cut an incline driven 115 feet, from which other drifts bisected the orebody. As yet no hoist had been installed, the ore being trammed by hand up the incline and shot directly from mine entrance through the Blake crusher to furnace shed below. The sheer volume of ore, and the ease with which it was mined, taxed the single water-jacket furnace and another Rankin-Brayton smelter was added late in 1880. Production soared to nearly 20,000 pounds of copper per day, providing fuel and water remained constant.

The prerequisites to successful mining were limited, however. Oak and manzanita along canyon floors, as well as mesquite on hillsides above camp, were rapidly chopped down by Mexican woodcutters; the cord wood transported by mule teams to fuel the steam engines, or converted to charcoal for the smelter. With furnaces running 24 hours a day the arc of destruction expanded with each passing week. The water supply, essential for furnace and crusher operation fluctuated with the seasons. During winter and spring the supply was adequate, but with coming of summer water flow slowed to a trickle, curtailing operation. To avoid costly interruption, Copper Queen management planned to construct a water line from the San Pedro River, but that venture was abandoned when water was struck on the 300-foot level of the mine.

There seemed no end to the Copper Queen orebody and the mine poured forth its wealth in unbelievable quantity. Its smelter ran continously without shutdown for repairs, belching smoke by day and casting an orange glow by night. The smelter meant business and a haphazard collection of tents and cabins, called Bisbee, sprung up around the Copper Queen works.

Despite the obvious richness of the Copper Queen Mine — it produced five million pounds of copper in 18 months — its owners felt uneasy. John Ballard and William Martin were experts at heavy construction, not miners. And while Edward Reilly had mining experience, he was suspicious of ore associated with limestone. All three men saw a glaring peculiarity in their mine: the deposit was nothing more than a giant plume thrusting through the limestone. It could not last forever. No doubt other ore lenses existed, but exploration was costly and deeper zones of mineralization might not be as rich as ore near the surface. The owners felt it would be prudent to ready the property for sale while there was still considerable ore reserve.

On the counsel of Louis Zechendorf and Albert Steinfeld the Copper Queen Mining Company was incorporated on April 2, 1881, under laws of New York state, and 250,000 shares of capital stock issued with a value of ten dollars per share. Officers of the newly formed corporation were: A. A. Hayes, Jr., of New York, president; Louis Zechendorf, treasurer and secretary; the executive committee consisted of Hayes,

Zeckendorf, and Albert A. Levi. The board of trustees were composed of the above men along with Charles and William Martin, John Ballard, and Edward Reilly. A prospectus entitled *A Description of Property of the Copper Queen Mining Company* was printed towards the end of 1881 for the purpose of attracting potential buyers, principally from English and French mining circles.

The prospectus touted the fact that between September 1880 and March 1881 more than two million pounds of black copper had been sent from the Mule Mountains to San Francisco, where it was loaded aboard Pacific Mail steamships and transported around the Horn to eastern refineries.

The Copper Queen prospectus was favorable but the $1,100,000 price placed on the mine was excessive. Although profitable, the Copper Queen could not be sold in Europe. Reports from American companies processing the mine's black copper, such as Pope, Cole & Company, the Revere Copper Company, the Ansonia Brass and Copper Company, and the Chemical Copper Company, drew attention of Anglo-American capitalists to the property. Transactions with the latter two firms ultimately placed the Copper Queen into the hands of James Douglas, and the proprietors of Phelps, Dodge & Company, William E. Dodge and D. Willis James, Jr. They altered the course of mining in the Mule Mountains and assured the future of copper production in Arizona Territory. It's a success story often told.

When it came to applying the Rankin-Brayton smelter to Arizona's argentiferous lead, it was a story of failure; not from the inadequacies of the furnace, but from insufficiency of ore. Examples of these failures are found in Pinal County, close to the Gila River, and on the east side of the Chiricahua Mountains. The latter will be examined first.

About the time the Copper Queen Mine was being brought into production, in 1880, a 20-mile long mineral-bearing zone was discovered on the eastern slope of the Chiricahua Mountains running from Camp Bowie to beyond Cave Creek. There, prospecting had delineated two belts of mineralization: the first, iron outcrops carrying carbonate ore with galena; the other was a quartz capped limestone with some lead rich in silver sulphuret. Within weeks the entire length of these mineral belts had been claimed by prospectors, and by year's end eastern capitalists, with visions of another Eureka or Leadville, had grabbed up the richest claims. The result was formation of the Califor-

The Rankin-Brayton furnace for argentiferous lead ore. The smelter's water-jacket was made of wrought iron in two or four sections, according to size, so that any of them could be removed without disturbing the rest of the furnace, making all parts accessible for cleaning and repairs. The lead well into which molten metal ran, was located outside the jacket. It was kept hot by a small fire, the dross and other impurities skimmed off before the bullion metal was ladled into molds. Located in the jacket, and surrounded by circulating water, the tuyeres never burned out. The jackets were oval in form with round corners, which prevented accumulation of slag, which tended to bridge the ore and prevent it from coming in contact with the blast. This furnace was made in 10, 20, 30, and 40 ton capacity models. (*Mining and Scientific Press*, March 14, 1891)

nia Mining District. Among these easterners were Pennsylvanians who had participated in the oil boom at Titusville, and were looking to invest their money in mining ventures in the Far West. John H. Galey was such a man.

Galey was a financial success when he arrived on the Arizona scene in 1880. He was born on February 4, 1840, near Cherry Run Camp Meeting Grounds, about 15 miles east of Parker, Pennsylvania; his father, Robert Galey (Gealy), having emigrated to the United States from Ireland in 1818. A companion of his on that voyage was Thomas Mellon, who one day would establish a financial dynasty. Under his father's tutelage John Galey developed into a shrewd investor, garnering big money at Titusville and other western Pennsylvania oil fields. When the oil boom faltered he saw profit in the precious metal excitement in southeastern Arizona.

In 1880 Galey bought some claims in the Turkey Creek area of the Chiricahua Mountains. When prospecting suggested rich bodies of silver ore, Galey went to Pennsylvania and encouraged friends (the McKinney brothers and the Painters[10]) to invest heavily in the project. He then returned to Arizona and put things on a fast schedule, developing his principal claim, the Texas Mine, and laying out a townsite named of course, Galeyville. Somewhere along the way he struck up an acquaintance with Samuel Wessels, a knowledgeable metallurgist who had been in charge of the Saturn Smelter in Salt Lake Valley in the early 1870s. Together, they laid plans to develop a number of claims into paying mines. Key to their operation was establishment of a smelter which would process not only ore from their properties but the product of adjacent mines and perhaps bullion smuggled across the border by Mexican miners.

Since much of their operation would be custom smelting, Galey and Wessels planned well. They choose a Rankin-Brayton smelter, an inexpensive furnace that spelled success for the Copper Queen Mining Company, 75 miles away. The decision made, John Galey went to San Francisco at end of November 1880 and purchased a 30-ton furnace from the Pacific Iron Works. By end of April 1881 the smelter had been shipped over the Southern Pacific Railroad to San Simon station, where it was unloaded, and hauled by teamsters to Galeyville over a newly blazed 25-mile long road across San Simon Valley.[11]

The furnace was placed on the Chiricahua Chief claim, atop a well-constructed foundation. Within 30 days a frame shed was fabricated around it, the Blake crusher, steam engines and blower installed. Fifty thousand bushels of charcoal was burned and stockpiled, as was ore and flux, the latter obtained from the Granite Gap claim, a property bonded that spring by John Galey for $15,000, 25 miles away in the central Peloncillo Mountains of the San Simon District of New Mexico. It was apparent he and Wessels had taken every precaution to assure successful smelter runs. With all bases covered, the furnace was blown-in on June 2, 1881. The first seven-day run produced 10 tons of bullion, which was shipped for refining to Pittsburgh, Pennsylvania. As with every smelting operation in the West, this activity created a sizeable community.

Over 100 men were on Galey's payroll by end of 1881. Other mining and milling enterprises, together with associated businesses such as freighting, blacksmithing, and merchandising, boasted Galeyville's population by several hundred more. Including a few stray cowboys, there were enough people to support restaurants, two hotels, an estimated eleven saloons, and a newspaper. Initial propaganda gave the impression that Galeyville would dwarf Tombstone. Geological quirks prevented that.

The smelter ran intermittently for six months, being shut down in early December 1881. It was fired up again for an 11 day run in late February or early March 1882. Shut down again in mid-March because of lack of ore, the *Arizona Daily Star* reported it would "resume operations at an early date."[12] It never did. The record of Galey's smelter was dismal; output for 1881-82 being estimated at 80,000 pounds of lead and 13,717 ounces of silver, total value $19,420. Why such a small return?

Some say production lagged because of fluxing problems and lack of ore. Others believe Galey's smelter was a scheme to bilk investors. The latter speculation is without foundation. No one bonds claims and purchase a new 30-ton furnace, and ship it 800 miles to deceive investors. Such actions indicate that Galey believed that silver-lead ore existed in large quantities on the eastern flanks of the Chiricahua Mountains, and that it could be more profitably reduced by smelting than by amalgamation. Galey recruited the talent to put such a plan into operation. Samuel Wessels had at least a decade of experience in the smelting business at several of the West's most successful smelters. The ore of the California Mining District was identical to that handled in the Salt Lake Valley of Utah. A trained metallurgist who understood the chemistry of slags, Wessels knew what types of ores to mix to maintain a proper charge for his furnace; he knew that before the furnace was blown-in.[13]

Galey's smelter failed because of other reasons. The most obvious being that there were no great fissures of argentiferous lead on the eastern flanks of the Chiricahua Mountains as at Eureka, Nevada, and Little Cottonwood Canyon, Utah. Only small replacement deposits, veins and stringers, without much depth, existed at Galeyville; core drilling would have told the story. While diamond drilling was known in 1880, it was costly and not widely practiced. At that time miners merely gambled that veins would open into bonanzas.

Keep in mind also that Galey's furnace had a voracious appetite. A 30-ton capacity meant it smelted 30 tons of ore every 24 hours. An efficient, trouble-free furnace, Galey's Rankin-Brayton smelter would have gone through the stockpiled ore in less than 100 days. Without a constant supply of ore smelting could not be profitably carried on. Then too, cost of freighting and fuel was high, as were Southern Pacific Railroad rates for shipping bullion. Add in the decline in the price of both lead and silver, and you have a scenario for failure.

If lack of ore was not enough to make Galey and Wessels tear their hair, iron ore, a necessary ingredient for silver-lead smelting, was not known in Arizona at that time. When Galey bonded the Granite Gap Mine he gambled its ore would be rich enough in iron to work as a flux. But his bond, of six months duration, ran out. Unable to raise funds enough to purchase the mine, Galey lost his source of flux. This was a common occurrence.

Several years later the proprietors of the Benson Smelting and Reduction Company faced a similar situation. That company started out using fluxing ore produced near Crittenden. The deposit contained only 500 tons of good iron, not enough to run a smelter, and the operation shut down. Vainly the company sought other sources of flux. Ore from other mines were tried; it was either magnetic or contained too much silicate to be useful. Finally owners of the Benson smelter located a mine near Lordsburg, New Mexico, which contained good fluxing ore which allowed them to run until 1885.[14]

The good times ceased at Galeyville when the cash dried up and the flux was not forthcoming. The furnace was sold, dismantled and moved elsewhere, some say to Benson where it was blown-in on December 28, 1882, as part of the Benson Mining and Smelting Com-

An 1882 view of the Copper Prince smelter at Bisbee.

pany, which processed silver-bearing lead ore from the Empire Mining District and mining localities as far away as Chloride in the northwest corner of the territory.[15]

It is doubtful John Galey's furnace went to Benson, for reason that this custom smelter, established by Munroe Salisbury, co-owner of the Gilmer-Salisbury stage line which ran throughout Nevada and Utah, was equipped with a 50 ton capacity water-jacket furnace manufactured by Fraser and Chalmers of Chicago. A second furnace of 80 ton capacity, by the same manufacturer, was installed in 1883.[16] More likely, Galey's furnace, which was a Rankin-Brayton 30 ton unit, went to one of two places. It may have been purchased by John A. Church of the Tombstone Milling & Mining Company to process manganese-silver ores of the Lucky Cuss and Luck Sure mines, and lead ore from the Hendricks and Atlanta claims at Bisbee. Or Galey's smelter may have been purchased by Frederick A. Tritle and John R. James, and transferred to the Mule Mountains to process ore from their Copper Prince claim. If that was the case, the furnace was for several years under the supervision of Frederick W. Heyne.[17]

What happened to John Galey and Samuel Wessels? Sam, who was the brother-in-law of Joseph Bowyer, another prominent resident of

Galeyville, may have hung around the area for a few years before disappearing from the historical record. John Galey, however, went on to fame and fortune. In spring 1882 he sold his interest in the Texas Mine and flew the coop. Supposedly he lost a fortune in this mining venture; his Pennsylvania friends may have also lost considerable money. According to newspaper reports Galey went to New York to make arrangements to pay off the company's indebtedness.[18] Which he probably did, for John Galey's career over the next 20 years is one of a gambler interested in wresting wealth from the earth, not that of a man seeking to bilk friends in fraudulent mining schemes.

Surviving his Arizona experiences, Galey went back to wildcatting. In 1892 he opened the great Mid-Continent oil field at Neodesha, Kansas, and drilled the fabulous Matthews gusher near Pittsburgh, Pennsylvania, which opened the McDonald oil pool. In 1901 he was instrumental in putting Texas on the map as an oil producing state by drilling the famous Lucas Gusher at Spindletop, near Beaumont on the Gulf Coast. Flowing at the rate of more than 100,000 barrels of oil per day, this gusher was the largest well in world history until capped several weeks later. The Mellons paid Galey a fortune for that property. John Galey went on to bring in wells at Coalinga, California, in what proved to be one of the richest fields in the United States, a discovery he could not interest the Mellons in. He was not successful in every oil venture, however. Galey sunk test wells in New Mexico, and lost his shirt in a dry hole at Tampico, Mexico.

During his up and down life, oil did not entirely replace precious metals in Galey's mind, for he dropped a lot of money trying to develop a gold mine in the Seven Devils region of northwestern Idaho, and he had gold claims in Alaska. His final mining attempt was at Joplin, Missouri, in 1916-17, when he reopened an old zinc property. Although he had "indifferent success" with that venture Galey lived in comfortable circumstances at Joplin until his death on April 12, 1918. His wife, sons and relatives returned his body to Beaver, Pennsylvania, where he had built an estate following his Texas oil success. He rests near a huge glacial boulder in the beautiful Beaver Cemetery.[19]

Bisbee and Galeyville were not the only places in Arizona Territory where Rankin-Brayton furnaces made history. There were several installed along the Gila River, east of Florence, in the early 1880s to process both silver and copper ores mined in the Mineral

Hill District, north of the Gila River, fifteen miles southeast of Pinal City.

Deposits of argentiferous lead ores and copper carbonate had attracted prospectors to the Mineral Hill District as early as 1870, when high grade ore was shipped to Yuma and transported to San Francisco. In 1874 an attempt was made to smelt the local product, the *Prescott Weekly Miner* noting on July 3 that a smelting furnace and assay office were the "last additions to the town of Florence." Built after the "Mexican style," the furnace smelted Silver Queen ore with "very good results."[20] The search for ore was spurred by development of the Silver King Mine after 1875; the mountains between Pinal City and the Gila River being combed for mineral wealth.

A Mr. Sands located silver-bearing ore on Mineral Creek in 1877. Forming the Mineral Creek Mining Company in October 1880, he employed 40 men to open a mine and grade roads to a stamp mill erected at a point on the Gila River called Riverside. Others followed in his footsteps. A Mr. Boulanger, "an old copper mining man" from the Lake Superior region nosing about Mineral Creek and Mineral Hill, was impressed by deposits that carried as much as 40 percent copper. The *Mining and Scientific Press* noted he "had not seen anything comparable to this new copper region."[21] Others apparently felt the same way and events transpired rapidly thereafter.

The Pinal Copper Company was organized in fall of 1880 to develop the copper potential on Mineral Creek. Incorporated in California on December 22, 1880, with offices in San Francisco, the company purchased a 30-ton Rankin-Brayton Furnace identical to that of the Copper Queen Mine at Bisbee. By early February 1881 the equipment was on the ground, and a Mr. Williams, "a practical smelter," had arrived to take charge of the erection of the furnace. This was "Uncle" John Williams, of Utah and Nevada fame, and one of the inventors of the Rankin-Brayton Furnace.[22]

Securing a number of claims, among them the Monroe, Ray, Esperanza, National, Tibbetts, Ida Bell, Green Mountain, and Reed, the Pinal Copper Company began operations in earnest about February 1, 1881. Fifty to 80 men were hired to open the mineral deposits, blaze eight miles of roads from the mines to the smelter site at Walnut Grove on the Gila River, and to stockpile ore and charcoal. The furnace was on the ground in early February, and 90 days later it was ready to run. Blown-in the first week of May 1881, the initial run of four days resulted in $2^1/_2$ tons of black copper.[23]

While the Pinal Copper Company was pursuing its dream, Aaron Mason, superintendent of the Silver King Mine,[24] eyeing the silver potential in the Mineral Hill District, organized in San Francisco the Pinal Consolidated Mining Company. In spring of 1881 he and his associates — Henry Donnelly, ex-superintendent of the Eureka Consolidated, Henry Linkton, F. Logan, S. Hart, and H. H. Noble — purchased two contiguous claims situated 10 miles south of Pinal City, the Silver Bell Quartz (for $20,000) and the Columbia (for $10,000).[25]

The company blazed a six mile long wagon road from the mines to a mill site on the Gila River. Work on a quartz mill began August, 15, 1881, as did development of the mines. The Pinal Consolidated Company also purchased ore from neighboring claims, which gave it a lot of both free-milling and smelting ore. So much of the latter did the company have that in early January 1882 Mason contracted with the Pacific Iron Works for a 40 ton galena silver plant, which was installed in May at a spot where the Gila River makes a U-shaped bend about eleven miles downstream from where the town of Kelvin would develop.[26]

Charcoal for the smelter was burned from wood harvested in the Pinal range, which separated the Globe and Mineral Creek areas on the east. Apache raiding in late 1881, however, disrupted deliveries of fuel from that source. Desiring a more reliable, and closer source of charcoal, in late 1881 or early '82 the company built five stone beehive kilns on a hillside close to its smelter.[27]

Fueled with charcoal produced from cottonwood and willow harvested from the river banks and hillsides, the smelter turned out 1,800,000 pounds of bullion during the first six months of 1883, which created a small town about the smelter, called Butte. Ultimately silver-bearing ore played out, the price of precious metal declined, transportation was expensive, and the enterprise failed; a story not unlike that of Galeyville, only the production was better.[28]

Known today as the "Cochran coke ovens," the five beehive kilns still stand as a testament to Aaron Mason's Pinal Consolidated Company. They may well have been the last charcoal kilns built in the West. Outside of some surface trash there is no evidence of the smelter or settlement.

The Cochran charcoal kilns built in 1881-82 by Aaron Mason on the north side of the Gila River. (Courtesy Arizona Historical Society)

With exception of the Copper Queen smelter, these examples of Rankin-Brayton furnace installations are typical of dozens of smelters erected throughout the American West. All were erected to process rich, shallow orebodies. When the carbonate ore vanished, leaned in tenor, or changed to sulphide, they disappeared. The Copper Queen smelter was an anomaly. Its orebody was immense. When played out, other lenses of copper were discovered in the nick of time. The smelter was enlarged, modified, and moved to more spacious surroundings at Bisbee. In 1902 it was torn down, replaced by a giant, modern facility 28 miles away at Douglas.

Bisbee demonstrated the virtues of the water-jacket furnace better than any locality in the West. The Copper Queen Mine was the best advertisement that a machinery manufacture could have, and the Pacific Iron Works capitalized on it by publishing endorsements in newspapers throughout the West. In spring it reproduced this letter (dated January 3, 1882) from Lewis Williams, superintendent of the Copper Queen smelter:

> In answer to your inquiries for some particulars in regard to the operation of your furnaces at the Copper Queen mine, [I] would say, that we have had one thirty-ton smelter in blast almost constantly since the 20th of August, 1880. The second smelter, of same capacity, was put in blast on the 28th day of August last, and has been running ever since, without an hour's intermission. We put through something over thirty tons of ore per day to each smelter, exclusive of flux, and our average bullion product from the two furnaces is from 25,000 to 27,000 pounds per day, running from 96$^{5}/_{10}$ to 98 fine. Our record for the year 1881 shows a production of nearly 5,000,000 pounds [of] bullion — the product of one smelter running continuously and one only four months of the year.
>
> On account of the scarcity of timber for making charcoal in our locality we use coke for fuel — mostly English — though we have occasionally used Trinidad in connection with it.
>
> To illustrate the facility with which these furnaces can be managed [I] would state that in several instances, when our coke supply has become short, I have banked the fires for several days at a time and started the stream of bullion again in fifteen minutes after putting on the blast. The furnaces are adapted to any classs of copper ores, are simple in construction, easily handled, and with proper care are always in order.
>
> [I] have had more than twenty years experience in copper smelting and have never seen a furnace that can compare with yours for reliable, effective and economical work; in fact I know of no other furnace now made that will handle successfully the copper ores of the western States and territories. The general character and details of the smelting plants furnished by your firm, I may also say, are well adapted to the business and fully meet all requirements.
>
> I make this statement in the general interest of the smelting business of the country and as due to your energy and enterprise in working up to such a state of perfection this important branch of mining machinery.[29]

Lewis Williams' sentiments about the versatility, adaptability, and reliability of Rankin-Brayton furnaces were echoed by mining men throughout the West, and by end of 1882 the Pacific Iron Works was the "headquarters for smelting equipment."[30] Sixty-one of their furnaces were working in Western mining camps.

So successful were Pacific's water-jacket furnaces and other mining equipment that the company set up distributorships throughout the country. In 1882 George M. Scott Company, a Salt Lake City mining machinery dealer, was made an agent for Pacific furnaces in Utah, Idaho, and Montana.[31] And A. P. Brayton went East to establish sales offices. In late 1882 he concluded an "arrangement" with the Excelsior Iron Works, a pioneer Chicago foundry, for construction of hoisting, milling, and smelting equipment. At the same time, the company opened a New York City office at 35 Broadway, under the charge of H. B. Murray.[32] Eastern and Midwest mining and smelting enterprises could now secure improved Pacific Coast machinery and save on freight charges.

Deciding to penetrate the small metal producer market, Pacific down-sized its 30-ton copper furnaces in 1883. Ten-ton furnaces were shipped to Battle Mountain, Nevada, and to Chihuahua and Sonora. The 30-ton units were still manufactured, however. In 1884 two went to Australia, and another one was set up at the Copperosity Mines near Gila Bend, Arizona. About this time mining companies began demanding larger units, a sign that ore grade and values were changing. A 60-ton plant was constructed for the New Jersey and Sonora Copper Company at Nacozari, Sonora.[33]

Only six water-jacket furnaces were produced in 1885: three for Mexico and three for Australia. With slacking of smelter sales, Pacific focused again on traditional milling equipment, most of which went to Arizona mining companies. Demonetization brought a decline in silver mining and the Pacific Iron Works turned its attention to manufacturing gold mills.

The Pacific Furnace was such a wonderful piece of machinery that other manufacturers copied its design. Within two years o its introduction, the Vulcan Iron Works of Bradford, Yorkshire, England, was producing exact replicas of the Pacific Furnace. Called the "Stewart Rapid Smelting Plant," this water-jacket furnace was introduced about 1884 in Australia at Cobar, New South Wales, to smelt pyritic ores. A few years later W. T. Garratt & Company of San Francisco marketed portable furnaces with many Williams' improvements. Two types of furnaces were made: a superheating furnace for copper smelting and a smaller galena furnace.

The superheating furnace had some novel departures. Its water-jacket extended all the way to the bottom of the furnace. Above the water-jacket was placed a superheater, no more than a jacket through which hot air was ducted. Working on the same principle employed in iron smelting, the superheater raised furnace temperature 300 to 500 degrees. The saving in fuel was said to be two percent for each 100 degrees of heat added to the air, in all from 6 to 10 percent, a very important item in the cost of smelting.

Garratt's furnace had other features: an adjustable bottom plate which could be raised or lowered to suit the nature of ore worked. The plate could also be removed to allow substitution of a lead-smelting bottom. A lead well siphon could be added in lieu of the "copper bottom." Tuyere boxes had closing lids so the blast could be shut off from any tuyere in case of clogging. And the furnace was not mounted on a flat plate in the usual manner, but was supported on four legs or pillars made of wrought iron pipe, screwed into brackets, riveted to the outer shell. In transporting the furnace, the legs, bottom frame, and spouts were removed to lesson weight.

Garratt made smaller lead furnaces in various sizes from 16 to 20 inches in diameter, and they too had adjustable bottom plates. Made of thinner iron, the small furnaces were sectional to allow for pack mule transport. These furnaces had four tuyeres connected to an annular wind-box in the usual manner but with flexible pipes and detachable nozzles, so the tuyeres could be instantly bared for examination. With four arches, every part of the interior was accessible for cleaning. Two of the arches were cut through at the bottom to permit a lead well and siphon to be used if required. Blast required for these furnaces could be provided by any means, including hand actuated bellows. These furnaces could smelt from one to five tons in 10 hours, according to amount of blast and nature of the ore. They were furnished with all required fittings.[34]

The Risdon Iron Works, founded about 1852 by Lewis Coffey and J. N. Risdon, also manufactured smelters. Skill with wrought iron brought this boiler making company lucrative contracts, and thrust it into the ranks of fabricators of smelter components early on. As pointed out in the first chapter, in 1865 Nathaniel Haskell employed Coffey and Risdon to create a wrought iron water-jacket for a copper smelter at Buchanan Hollow in Fresno County. This was the first water-jacket to be applied to an American furnace. For several years thereafter, Risdon and Coffey was sales agent for the Haskell furnace, building units which were used in Calaveras and Fresno counties, and at Williams' Fork on the Colorado River, and in Nevada.[35]

FIG. 1.

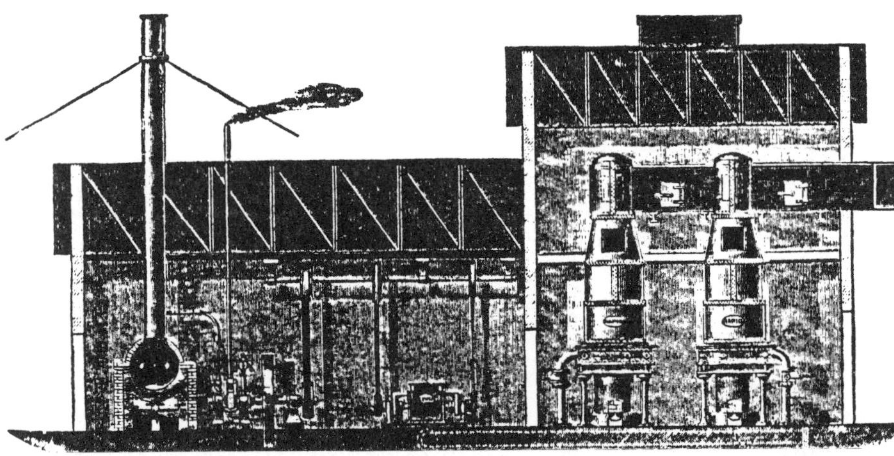

Above and right: "Stewart's Rapid Smelting Plant," manufactured in England, was identical to the Pacific Furnace. Produced in a number of sizes, the one shown was five feet in diameter, and had eight five-inch tuyeres. The water-jacket extended from the tuyeres to the charging door. The water-jacket required as coolant 450 gallons of water per hour. The blast was supplied by a Roots No. 5 blower running at 150 revolutions per minute. One steam engine powered both blower and crusher. (*Scientific American*, Supplement No. 495, June 27, 1885)

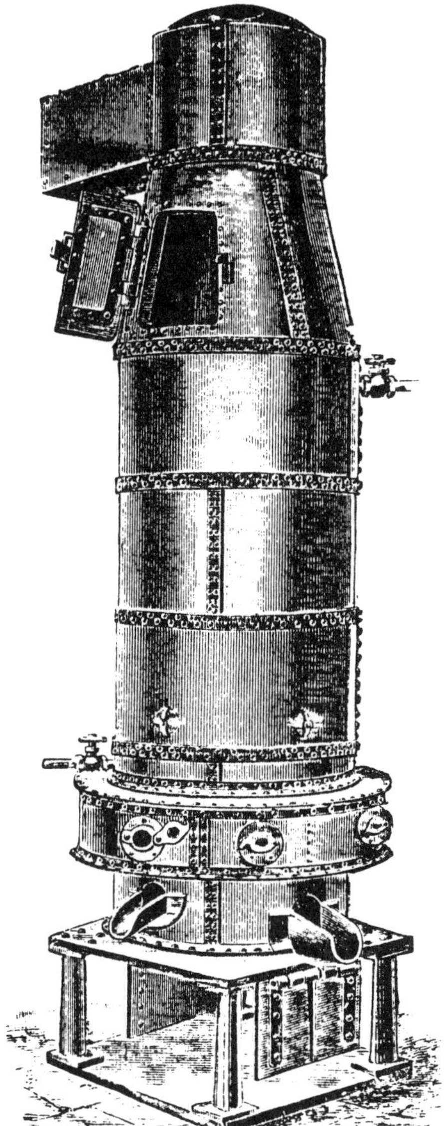

Fig. 1.

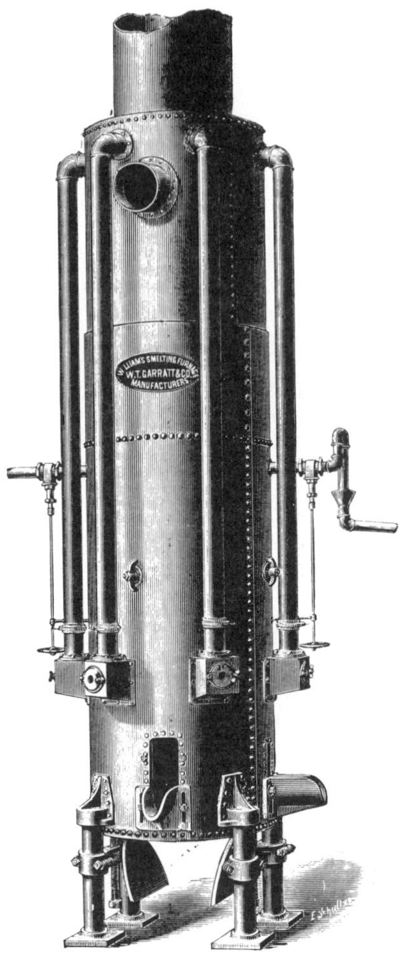

Fig. 2.

The superheating furnace, arranged for copper smelting, as built by W. T. Garratt & Company of San Francisco. Above the water-jacket was placed a superheater (*AA*) exposed to hot gases inside and connected to the tuyeres by six dam pipes as shown. Air was forced in at the large nozzle (Fig. 1). (*Mining and Scientific Press,* June 30, 1888)

By 1872 the company had changed its name to the Risdon Iron and Locomotive Works, and its location to the corner of Howard and Beale streets. With a work force of skilled foundrymen and mechanics, manning up-to-date equipment, it had the capacity to build just about anything, and it was not long before the company was producing small, sectional water-jacket silver-lead smelting plants similar to those built by the Pacific Iron Works.

Equipped with Baker blowers, and an improved Corliss engine and boiler, these small smelters worked well with carbonate ores of lead and copper, but when tenor of ore changed, metal percentages dropped, or contained copper pyrites and glance, smelting became more complex, calling for special designed large capacity furnaces.

Among the finest equipped foundries in the nation, the Risdon Iron Works in the 1880s and '90s was frequently called upon to build galena-silver smelting furnaces, in a variety of sizes from 36 by 42 inches, up to 36 by 78 inches. Larger sizes were built on request. These water-jacket Raschette smelters had a unique feature: they were built in sections so that they could be extended to increase capacity as mine output demanded. Many of these smelters were adopted by the best Eastern reduction works, and gave excellent results.[36]

As soon as East and West was linked by transcontinental railroads Midwest foundries got into the act of building smelting plants, the most active being Fraser & Chalmers, a company that had its inception in Chicago following the great fire of 1871. A Scot emigrant, Thomas Stuart Chalmers, was the driving force in this company. It is said his life is the story of Chicago's industrial development.

Thomas Stuart Chalmers was born June 14, 1816, at Dronley, near Dundee, Scotland. His father, William Chalmers, being a manufacturer of linen. Thomas attended school until he was 14 years of age, then became an apprentice machinist at Brown & Allen, Dundee manufacturers of spinning machinery. At that time the manufacture of steam-spinning machinery was in its infancy.

At termination of his apprenticeship, seven years later, Chalmers went off as a journeyman machinst to a place called Fyfe in Fyfeshire, about 15 or 20 miles from Dundee, where Brown & Allen had a plant. For three years he worked as a machinist putting up power spinning and carding machines in various mills throughout Scotland.

In 1843 Thomas and his wife, Janet Telfer, emigrated to the United States, settling on a farm site between Joliet and Lockport,

Illinois. Finding it difficult to make a living tilling soil, Thomas walked the 40 miles into Chicago in spring of 1844 and sought employment as a machinist. His first position was in the Nickerson machine shop at the corner of Randolph Street and Fifth Avenue. Chalmers worked through the summer doing all kinds of repair work at a dollar a day. Nickerson was unable to collect from his customers and Chalmers received little pay, which forced his return to the Lockport farm in the summer.

Returning to Chicago in fall of 1844, Thomas secured a job as chief engineer on the *Lady of the Lake*, a passenger vessel running between Chicago and Buffalo. He served on the boat a whole season receiving $40 per month and board; the first money he had seen in America. Concluding that job he returned to Chicago and went to work for P. W. Gates and Hiram Scoville, who had a blacksmith shop on West Water Street. They were both blacksmiths, Chalmers being the only machinist there. The work performed by the three men was typical of blacksmith shops of the time; they worked on anything that came along, repairing farm equipment, building wagons and plows. They built Chicago's first water works, a single pump at the foot of the river. After that the company grew rapidly.

In 1855, with P. W. Gates, E. S. Warner, A. G. Warner and David R. Fraser, Thomas Chalmers formed the firm of Gates, Warner, Chalmers & Fraser, which was succeeded in 1857 by the Eagle Works Manufacturing Company. Chalmers was associated with Gates until the great fire of 1871. After that catastrophe Gates liquidated the corporation, and in winter of 1871-72 the firm of Fraser & Chalmers was formed with Chalmers, Fraser, Robert Anthony and David Marchant being stockholders.

Fraser & Chalmers commenced operations with 60 men in a 60 x 145 feet foundry. Besides doing general repair work, the company became the largest builder of sawmill machinery in the United States. At the same time it fabricated mining equipment, following patterns designed by San Francisco foundries. For several decades the company produced a variety of stamp mills, hoists, and steam engines for copper companies of the Great Lakes region, and emerging lead mines in Missouri and Illinois. Mining companies of the Rocky Mountain region also purchased Fraser & Chalmers equipment. Although Thomas Chalmers retired from active business in 1893, the company continued as a major source of American mining equipment. When it was combined in 1901 with the Milwaukee firm of A. P. Allis, builders of hydraulic equipment, the company was working 1,500 men in Chicago and 1,000 men at shops in Frith, Kent, England.

Thomas Stuart Chalmers died at his Chicago residence on July 13, 1903. He was survived by his widow Janet (whom he married on July 2, 1840, in Greenock, Scotland), and five children. His life is the story of American industrialism. A staunch Republican he supported the Union cause during the Civil War. His humanitarian views drew to him such men as John Wentworth, Henry M. Singer, William B. Ogden, and Allan Pinkerton, with whom Chalmers was associated in the underground railway for escaping slaves.[37] Now back to Chalmers' role in Western smelting.

In 1880 Fraser & Chalmers designed and built its first water-jacketed smelting furnace. That story started one frosty morning early in March of that year when Horace Brown of the Hecla Consolidated Mining Company of Glendale, Montana Territory, entered the company's machine shop office at the corner of Fulton and Union Streets. Dressed in miners' overalls, he carried a suitcase and a roll of roughly made pencil drawings.

Without much formality, Brown said he wanted designed and built a 36 x 60 inch lead-smelting furnace with water-jackets. Being before the advent of blue-printing, Charles C. Christensen, mechanical engineer and draftsman, was appointed to wait on Brown and make all the drawing to suit his ideas.

The general drawings were made on heavy, white paper, inked and colored to show the different kinds of material. A tracing was made of these drawings for the customer. The details were drawn on plain yellow paper, inked and traced. These tracings were pasted on heavy cardboards, varnished over and forwarded to the shops. The result was the furnace pictured on page 206.

The curb of this furnace was two feet four inches high, and had heavy, ribbed cast-iron sides made up in six sections and bolted together as shown. The cast-iron spout was bolted to the end curb plate under the tap opening. The water-jacketed portion of the furnace was 36 x 60 inches at the tuyere line and 48 inches high. It consisted of eight cast-iron side jackets, two rear-end jackets, all boshed $8^1/_2$ inches, and one straight, front-end jacket, 37 inches high, leaving the 11-inch high tapping breast. All the jackets had ears cast on for connection by clamps and keys, and were also provided with heavy T-steel binders.

To each jacket was bolted a cast-iron overflow spout with nipples

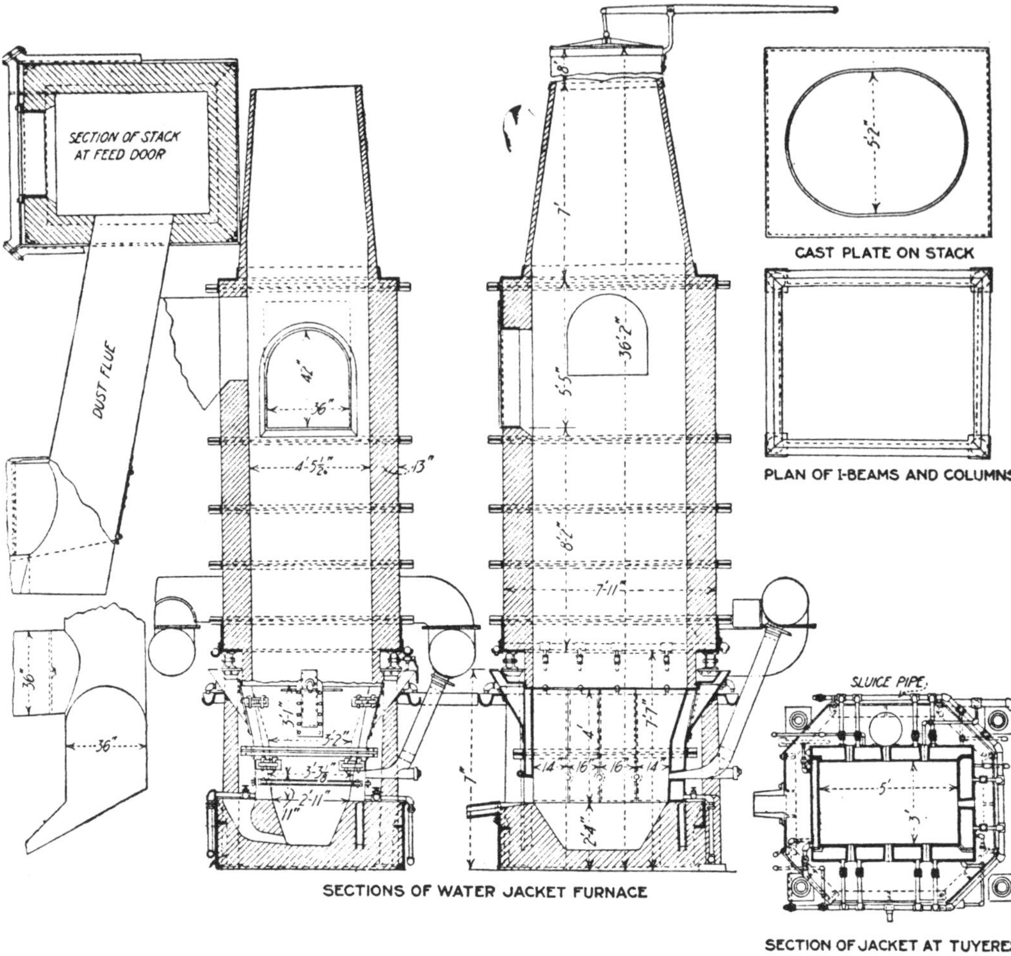

Diagram of the first water-jacketed lead-smelting furnace built by Fraser & Chalmers in 1880. It was made to order on short notice for the Hecla Consolidated Mining Company of Glendale, Montana Territory. (*Engineering and Mining Journal*, November 10, 1917)

By 1882 Fraser & Chalmers were manufacturing portable water-jacket furnaces identical to those designed by the Williams brothers. Called the Nichols smelting furnace, they were used extensively on the East Coast. (*Mining and Scientific Press,* August 25, 1883)

and elbow for discharge into the overflow launder surrounding the furnace. There were seven Y-type tuyeres $3^1/_2$ inches in diameter tapering to $2^3/_4$ inches in diameter in the jacket tuyere openings. They had canvas hose connections to the air gates which were connected by short cast-iron nozzles to the 20-inch bustle pipe. The sluice pipes were laid horizontally on top of the curb, bending over at the edge, and came down below the floor and there connected to a main discharge pipe running outside the furnace building.

The supply water main was placed right above the overflow spouts with a tee and nipple connecting to each spout. Hangers, supporting this water main, were bolted on the brick-stack deck plate. Four cast-iron columns, 7 feet long, 6 inches in outside diameter at top, 8 inches in outside diameter at bottom, of 1-inch metal, supported the mantel frame which consisted of two 7-inch 15-pound I-beams at the sides and ends, with corner angles, pipe spacers and bolts for connections.

Upon the mantel frame rested the heavy, cast-iron angle deck plate supporting the brick stack, the outside dimensions of which were 7 feet 11 inches by 6 feet $7^1/_2$ inches, with 13-inch thick brick walls. The binders were made of T-steel — five sets as shown — clamped together at the corners by U-bolts, washers and keys. to the cast-iron top plate was bolted a sheet-steel, conical hood, 7 feet high, which was brick-lined. On top of the hood was a steel stack, 8 feet high and 48 inches in diameter, with damper, also brick lined. The feed-door frame, 36 inches wide by 42 inches high, was of arched design. The downtake opening was 36 x 36 inches with circular top. A sheet-steel downtake of same dimensions, with damper, completed the furnace.

The general dimensions were as follows: Floor to top of curb, 2 feet 4 inches; to tuyere line 3 feet 3 inches; to top of jackets 6 feet 4 inches; to stack deck plate 7 feet 7 inches; to charging floor 15 feet 9 inches; to top of brick stack 21 feet 2 inches; to top of hood 28 feet 2 inches; and total height of furnace, with stack, 36 feet 2 inches. The furnace was designed, built and shipped in a remarkably short time to the Hecla Consolidated Mining Company of Glendale, Montana Territory. Capacity of the furnace was 50 tons of charge per day of 24 hours. A similar furnace was installed in 1883 at the Benson Smelting Company, in Arizona Territory. [38]

Fraser & Chalmers introduced a small circular water-jacket furnace in 1882. Called the Nichols Furnace, it was so identical to the Rankin Brayton smelter, that it undoubtedly was a spinoff of the Williams' design. Marketed until 1915, the Nichols Furnace had one novel accesssory, a removable outside well or hearth.

While not a new idea, this improvement introduced in 1883, was no more than a frame of wrought iron on wheels, surrounded by soft sheet steel between which was a water space. A smaller inner frame corresponded in size and height to the opening in the furnace. The well itself was built of ribbed cast plates, bolted together, lined with firebrick and mineral wood to prevent chilling, and mounted on wheels with a flat arched brick lined cover. Near the top, on the side away from the furnace, was a slag spout and below on one side a tap-hole and spout for the metal, which could be tapped at intervals, or in case of lead, a small outside dipping well could be bolted on to the furnace itself. The well was guided to its place against the furnace by two outside rails.

The connection was made by running the jacketed face of the well against the corresponding hole in the furnace, attaching rubber hose connections to the water supply pipes, and then letting the blast on to the furnace. Any slag which might leak between the two faces was immediately chilled, forming a seal between the two units. The advantages claimed were removal from the furnace itself of all brickwork which could be cut by the slag, so that repairs could be made without hindrance to the smelting operation. Discharge of furnace products was through an opening in the water-jacket itself, which was claimed to be practically indestructible. Discharge into the well, being below the level of the slag, prevented the escape of fumes, and kept the contents of the well uniformly hot.[39]

Manufactured in three sizes, the Nichols Furnace was introduced at the Laurel Hill Chemical Works, Laurel Hill, Long Island, where it was put to work smelting roasted pyritic ore containing between three and four percent copper. Results were beyond expectations, and the furnace was soon utilized to develop small rich properties in countries where transportation facilities were meager, as in Mexico, Peru, Chile, and other Latin-American republics.[40]

The late 1880s and early 1890s were hard years for all San Francisco foundries. No longer were Western mining districts their exclusive turf. Then too, the mining frontier was expanding. Eastern manufacturers could better serve Sonora, Chihuahua, and interior Mexico. San Francisco companies also had to contend with high freight charges and stiff tariffs. Rather than continue in what was perceived as a loosing struggle, some foundrymen choose to retire or change direction.

In May 1890, after 35 years with the Pacific Iron Works, A. P. Brayton left the firm for the Pelton Water Wheel Company. It was a harbinger of things to come. In late January 1892 the Pacific Iron Works was absorbed by the Union Iron Works, which retained the company's old facilities for construction of smelters, mills, and hoists, as well as a center for job and repair work.[41] Acquisition of the Pacific Iron Works gave Union a valuable line of machinery: the Pacific water-jacket furnace, the White-Howell ore furnace, and a variety of hoists, engines, and stamp mills.[42] From that time forward a great deal of Union's equipment resembled the old Pacific line.

With the world crying for copper at 1900, the Union Iron Works moved into the manufacture of gigantic smelting plants. With a capacity of 300 tons, the main feature of these smelters were their "adaptability." They were able, without preliminary roasting, to reduce sulphide ores to a matte suitable for bessemerizing. Patterned after the Raschette furnace, these smelters were immediately accepted by the mining world with installations at the Douglas smelter of the Copper Queen Consolidated Mining Company, and at Nacozari for the Greene Consolidated Copper Company and the Copete Copper Company. Others were installed in Arizona, Utah, Colorado, and Montana. We cannot conclude this book without saying something about the latter years of the men who perfected the portable water-jacket furnace, the Williams brothers.

Following his Pinal Copper Company experience, John, Jr., played a prominent role in development of copper mines at Globe. He also managed the Commercial Mining Company of Prescott, a property purchased by the Copper Queen Company. In summer of 1890 he installed a 30-ton Rankin-Brayton Furnace at that mine. It was his last smelter installation. He died at age 47, in San Diego on January 16, 1892. His body was returned to San Francisco for interment in a special railroad car.

His brothers Lewis and Ben, more renowned in the history of Arizona copper production, carried on in the management of the Copper Queen Consolidated Mining Company until after the turn of the century. In 1901 Ben and Lewis retired from Bisbee. Relocating to Milton near San Francisco, they supervised the working of the old Napoleon Mine, a gold, copper, and silver producer in Calaveras County. They did not give up their commercial interests in southeastern Arizona, however. Both were large stockholders in the Bank of Bisbee and the Bank of Douglas, institutions that Ben helped organize. Later the brothers moved to Los Angeles, where Lewis died on September 5, 1907. He was 73 years old. Ben lived on for quite a while. Although he had a palatial home at 1938 South Harvard Boulevard in Los Angeles, he died at an apartment at 3355 Wilshire Boulevard on September 1, 1925.

* * *

1. Lynn R. Bailey, *Supplying the Mining World. The Mining Equipment Manufacturers of San Francisco, 1850-1900.* Tucson: Westernlore Press, 1996, pp. 42-48.

2. This was not a new idea, for E. P. Ball of Chicopee, Massachusetts, had built a steam powered crushing mill in the late 1840s. A single stamp unit, the Ball Mill as it was called, was introduced into North Carolina gold mining. By 1857 the mill had evolved into a multiple stamp unit, widely used in the copper region of northern Michigan.

3. *Mining and Scientific Press*, March 30, 1867.

4. *Ibid.*, October 22, 1870, p. 88.

5. Issue of June 3, 1871.

6. *Mining and Scientific Press*, February 7, 1880, p. 89.

7. *Ibid.*, October 21, 1882.

8. Lynn R. Bailey, *Bisbee, Queen of the Copper Camps*. Tucson: Westernlore Press, 1983, pp. 21-23.

9. August Raht was born February 25, 1843, at Dillenburg, Duchy of Nassau, Germany, where his father was president of the Court of Appeals. After passing through the gymnasium, he studied mining and metallurgy, graduating from the Polytechnic Institute at Hesse-Cassel in 1863, and continued his studies at Freiberg. He came to the United States about 1867, taking a position with the Ducktown Smelting and Reduction Company at Ducktown, Tennessee. Except for a two year sojourn to Italy and Africa, Raht worked for this smelting company until he went West to built the smelter of the Neptune Mining Company. Completing that project, he took a position with the Mingo Smelter in Utah's Salt Lake Valley. In 1881 he supervised the building of the Francklyn Smelter of the Horn Silver Mining Company, also in Salt Lake Valley. When that smelter closed four years later, he went to Wickes, Montana, and later to Helena, where he built the plant of the Helena & Livingston Reduction Company, which was consolidated with the Great Falls plant built by Anton Eilers, forming the United States Smelting and Refining Company, of which Raht was manager.

August Raht was prominently connected with development of the smelting industry in the United States during its most important period, both in copper and lead, although more prominently in the latter. He originated many operational improvements and was one of the first of the Western metallurgists to recognize the importance of being able to treat complex sulphide ores, his work of this kind at the Philadelphia smeltery at Pueblo, Colorado, being largely the cause of its success. This plant was one of the pioneers in providing itself with equipment for treating lead sulphide ores on a large scale. Raht was one of the first to recognize the possibility and benefit of low fuel in blast furnace operations connected with lead smelting. In the early stage of lead smelting, and before the practice of converting mattes became common, he worked up copper byproducts at the Philadelphia plant into blister copper. During the greater part of his career he successfully devoted himself to solving the new problems which arose with evolution of lead smelting. August Raht died at the Palace Hotel in San Francisco, December 25, 1916. *Engineering & Mining Journal*, January 20, 1917, p. 164; February 24, 1917, p. 358; March 10, 1917, pp. 429-30.

10. There were seven McKinney brothers, but it is likely that only two, James and John, partnered with John Galey in mining development in the California Mining District in 1880-81. James C. McKinney, the most commercially active, was born at Pittsfield, Pennsylvania, in 1844. He joined a Pennsylvania railroad surveying corps in 1861, which ended at

Franklin two years later. Thereupon he started a lumber yard in Oil City, and opened a second yard in Franklin. In 1868, while living at Titusville, he married Agnes E. Moore of Franklin. With brother John L., James went from lumbering to oil production, gradually gaining control of 100,000 acres of oil-producing land. In 1889 they sold their rights to Standard Oil, taking management positions with the company. James had other interests, however. He was a director of the Titusville Iron Works, and was responsible for constructing a paved road from Titusville to Oil City, and he was a director of both the Commercial Bank and the Second National Bank. In 1918 he founded the Titusville Trust Company. A Jeffersonian Democrat, he was interested in politics, serving as mayor of Titusville from 1905 to 1908. James Curtis McKinney died in 1924.

The three Painter brothers, Augustus, Reuben and William, were originally from Olean, New York. Reuben and William relocated to Bradford, Pennsylvania, where they participated in the area's oil boom. Augustus became prominent in Pittsburgh iron production. In company with John H. Galey, the Painters purchased a half interest in the Lizzie Mine at Tombstone, as well as a share of the Vizina Mine, also at Tombstone. At the same time they bonded the Bell and the north extension of the Mono, in the Turquois District. The Painters were also interested in mining property at Oro Blanco in Pima County. In partnership with John Galey, the Painters purchased considerable mining property in the California Mining District on the eastern slopes of the Chiricahua Mountains, and were co-founders of the Galeyville townsite in mid-November 1880, along with John Galey, H. B. Maxson, and T. E. Sumner.

11. *Arizona Daily Star*, November 30, 1880; *Arizona Weekly Star*, April 31, 1881.

12. Issue of March 28, 1882.

13. *Tombstone Daily Nugget*, December 15, 1881.

14. *Daily Star*, January 12, 1883.

15. *Mining and Scientific Press*, January 19, 1884

16. According to the *Arizona Daily Star* of January 12, 1883, the first furnace, of 50-ton capacity, was blown-in under the charge of John Williams, "formerly in charge of the Horn Silver Mining Company of Utah;" none other than the inventor of the Rankin-Brayton Furnace.

17. *Daily Star*, January 12, 1883, p. 4; B. S. Butler, E. D. Wilson, and C. A. Rasor. *Geology and Ore Deposits of the Tombstone District, Arizona*. Tucson, University of Arizona, 1938, pp. 43-44.

18. *Arizona Daily Star*, March 28, 1882.

19. Lynn R. Bailey and Don Chaput, *Cochise County Stalwarts*. Tucson: Westernlore Press, 2000, Vol. I, pp. 135-37.

20. *Prescott Weekly Miner*, July 3, 1874.

21. *Mining and Scientific Press*, October 2, 9, 1880, pp. 213, 229.

22. *Pinal Drill*, September 11, October 2, 1880, p. 3; *Mining and Scientific Press*, December 4, 1880, p. 35; March 5, 1881, p. 152.

23. *Mining and Scientific Press*, May 14, 28, July 23, 1881.

24. A native of Massachusetts, Aaron J. Mason was also one of the developers of Pinal City. After pursuing silver mining and smelting north of the Gila, Aaron developed onyx deposits near Big Bug in 1890-91. In 1899 he joined his son Ernest at Bisbee, becoming proprietor of the Norton House, a hotel,.

25. Owners of Silver Bell Quartz were R. Conn, H. B. Murray, and W. Hunton. Owners of Columbia were Champion, Goodfellow, T. Farrel, W. Hapton and J. Hixon. *Mining and Scientific Press*, June 25, 1881, p. 425.

26. *Mining and Scientific Press*, January 14, 1882.

27. *Pinal Drill*, July 9, 1881, p. 4, September 17, 1881, p. 4; *Los Angeles Mining Review*, July 16, 1881, p. 37; October 1, 8, 1881, p. 221, 241, 244.

28. *Arizona Daily Star*, June 19, 1883; Harry Ellington Brook, "The Great Silver King Mine," *Los Angeles Mining Review*, Vol. IX, No. 12, March 23, 1901, p. 17.

29. Published in the *Arizona Daily Star*, March 15, 1882.

30. *Mining and Scientific Press*, January 14, 1882.

31. *Tribune*, September 8, 1882.

32. *Mining and Scientific Press*, March 4, 1882.

33. *Ibid.*, June 26, 1880; March 8, 1884.

34. For information regarding the "Stewart Rapid Smelting Plant" see *Scientific American,* Vol. XIX, Supplement No. 495, June 27, 1885, pp. 7900-01; W. T. Garratt's portable furnaces are detailed in the *Mining and Scientific Press,* June 30, July 7, 1888.

35. *Ibid.,* April 7, May 12, 1866.

36. *Ibid.,* March 5, 1892; January 7, February 11, 1893.

37. Thomas and Janet Chalmers had six children. Those that survived were: Mrs. F. D. Potts, of Marseilles, Illinois, Mrs. Christina Owens, William J. Chalmers, Mrs. William Dandy, and Thomas Chalmers, Jr., all of Chicago. For additional details on the life of Thomas Chalmers and the Fraser & Chalmers Company consult: *Mining and Scientific Press,* July 25, 1903, p. 59; August 1, 1903, p. 68; *Mines and Minerals,* August 1903, p. 22; and the *Engineering and Mining Journal,* July 18, 25, 1903, pp. 97-98, 133-34.

38. Charles C. Christensen, "An Early Type of Water-Jacketed Smelting Furnace," *Engineering and Mining Journal,* Vol. 104, No. 19, November 10, 1917, pp. 836-837.

39. *Mining and Scientific Press,* August 25, 1883.

40. Charles C. Christensen, "Designing Small Copper Smelting Plants," *Engineering and Mining Journal,* January 30, 1915, p. 225.

41. *Mining and Scientific Press,* January 29, 1887; May 31, 1890.

42. *Ibid.,* February 13, 1892.

BIBLIOGRAPHY

NEWSPAPERS AND PERIODICAL

Arizona Daily Star, Tucson
Bisbee Daily Review
Cassier's Magazine
Engineering and Mining Journal, New York
Eureka Sentinel, Eureka, Nevada
Iron Age
Los Angeles Evening Express
Los Angeles Mining Review
Los Angeles Times
Mines and Minerals
Mining and Scientific Press, San Francisco
Mining World
Mining Science
Pinal Drill
Prescott Weekly Miner
Salt Lake City Daily Herald
Salt Lake Daily Tribune
Salt Lake Mining Review
San Francisco Alta California
San Francisco Chronicle
Scientific American, New York City
Southern Utah Times
Tombstone Epitaph
Tombstone Daily Nugget
Tucson Citizen
Tucson Star

BOOKS

Angel, Myron. *History of Nevada.* Oakland: Thompson & West, 1881. Reprint by Howell-North, Berkeley, 1958.

Annual Report of the Commissioner of Patents. Washington, D.C. Government Printing Office, 1860-80.

Appleton's Cyclopaedia of Applied Mechanics. New York: D. Appleton & Company, 1889. Two volumes.

Arrington, Leonard J., and Gary B. Hansen. *"The Richest Hole on Earth." A History of the Bingham Copper Mine.* Logan Utah: Utah State University Press, 1963.

Bailey, Lynn R. *Bisbee, Queen of the Copper Camps.* Tucson: Westernlore Press, 1983.

———. *Old Reliable. A History of Bingham Canyon, Utah.* Tucson: Westernlore Press, 1988.

———. *A Tale of the "Unkilled." The Life, Times, and Writings of Wells W. Spicer.* Tucson: Westernlore Press, 1999.

———., and Don Chaput. *Cochise County Stalwarts, A Who's Who of the Territorial Years.* Tucson: Westernlore Press, 2000.

Bancroft, Hubert Howe. *History of California.* San Francisco: The History Company, 1890

———. *History of Nevada.* San Francisco: The History Company, 1890.

———. *Chronicles of the Builders.* San Francisco: The History Co., 1891. 7 vols.

Bowen, Eli. *The Pictorial Sketch-Book of Pennsylvania.* Philadelphia, 1852.

Butler, B. S., E. D. Wilson, and C. A. Rasor. *Geology and Ore Deposits of the Tombstone District, Arizona.* Tucson, University of Arizona, 1938.

Callender, J. F., and Jan C. Wilt, R. E. Clemons (eds). *Land of Cochise, Southeastern Arizona.* Albuquerque: New Mexico Geological Society, Inc., 1978.

Canty, J. Michael, and Michael N. Greeley (eds.). *History of Mining In Arizona.* Tucson: Mining Club of the Southwest Foundation, 1987.

Church, John A. *Notes of a Metallurgical Journey in Europe.* New York: D. Van Nostrand, 1873.

Crookes, William and Ernest Rohrig. *A Practical Treatise on Metallurgy.* London: Longmans, Green & Co., 1870. 3 volumes.

DeGroot, Henry. *The Comstock Papers.* Reno: Grace Dangberg Foundation, 1985.

Dictionary of American Biography. New Yor:k Charles Scribner's Sons, 1975. 10 volumes. Edied by Allen Johnson and Dumas Malone.

Dunning, Charles H., and Edward H. Peplow, Jr. *Rock to Riches.* Phoenix: Southwest Publishing Company, Inc., 1959.

Emmons, S. F., and G. F. Becke. *Statistics and Technology of the Precious Metals.* Washington: Government Printing Office, 1885.

Farish, Thomas E. *History of Arizona.* Phoenix: 1915. 7 vols.

Fell, James E., Jr. *Ores to Metals. The Rocky Mountain Smelting Industry.* Lincoln: University of Nebraska Press, 1979.

Geary, Edward A. *A History of Emery County.* Salt Lake City: Utah State Historical Society, 1996.

Hilton, George W. *American Narrow Gauge Railroads.* Stanford: Stanford University Press, 1990.

Hittel, John S. *The Commerce and Industry of the Pacific Coast of North America.* San Francisco: Bancroft and Company Publishing, 1882.

Iles, Malvern Wells. *Lead Smelting.* New York: J. Wiley & Sons, 1902.

Jackson, W. Turrentine. *Treasure Hill. Portrait of a Silver Mining Camp.* Tucson: University of Arizona Press, 1963.

Keller, Charles L. *The Lady in the Ore Bucket. A History of Settlement and Industry in the Tri-Canyon Area of the Wasatch Mountains.* Salt Lake City: University of Utah Press, 2001.

Lingenfelter, Richard E. *Death Valley & the Amargosa. A Land of Illusion.* Berkeley: University of California Press, 1986.

McNair, James B. *Simon Cameron's Adventure in Iron, 1837-1846.* Los Angeles, by author, 1949.

Marcosson, Isaac Frederick. *Metal Magic; the Story of the American Smelting and Refining Company.* New York: Farrar, Strauss, 1949.

Mead, Howard, Jill Dean, and Susan Smith. *Portrait of the Past. A Photographic Journey Through Wisconsin,* Vol. I (1865-1920). Madison: Wisconsin Tales & Trails, 1973.

Merriam, C. W. *Geology of the Cerro Gordo Mining District, Inyo County, California.* Geological Survey Professional Paper 408. Washington, D.C.: Government printing Office, 1963.

Molinelli, Lambert, and Company. *Eureka and Its Resources.* San Francisco: H. Keller and Company, 1879. University of Nevada Press reprint 1982.

Mowry, Sylvester. Arizona and Sonora: *The Geography, History, and Resources of the Silver Region of North America.* New York: Harper & Brothers Publishers, 1864.

North, Diane M. T. *Samuel Peter Heintzelman and the Sonora Exploring & Mining Company*. Tucson: University of Arizona Press, 1980.

Overman, Frederick. *A Treatise on Metallurgy; Comprising Mining, and General and Particular Metallurgical Operations.* New York: D. Appleton & Co., 1887.

Percy, John. *The Metallurgy of Lead, including Desilverization and Cupellation*. London: John Murray, 1870.

Percy, John. *Metallurgy: The Art of Extracting Metals from their Ores*. London: John Murray, 1875.

Peters, Edward Dyer, Jr. *Modern Copper Smelting*. New York: Engineering and Mining Journal, 1903.

Piggot, A. Snowden. *The Chemistry and Metallurgy of Copper.* Philadelphia: Lindsay and Blakiston, 1858

Reeder, Clarence A., Jr., *The History of Utah's Railroads 1869-1883*. New York: Arno Press, 1981.

Seegmiller, Janet Burton. *A History of Iron County*. Salt Lake City: Utah Historical Society, 1998.

Shelley, George F. Early *History of American Fork. With Some History of a Later Day.* American Fork: American Fork City, 1945.

Spence, Clark C. *British Investments and the American Mining Frontier, 1860-1901*. Ithaca: Cornell University Press, 1958.

Svedelius, G. *Hand-Book for Charcoal Burners.* New York: John Wiley & Son, 1875. Translated from the Swedish by R. B. Anderson.

U.S. *Senate Report No.* 2, 50th Congress, Second Session, 1888-89.

Wallace, William J., and Edith Wallace. *Digging into Death Valley's History. Three Studies in Historic Archaeology*. Ramona, California: Acoma Books, 1981.

Watt, Ronald G. *A History of Carbon County.* Salt Lake City: Utah State Historical Society, 1997.

White, Brooks. *Galeyville, Arizon Territory 1880. Its History and Historic Archaeology.* Raleigh, North Carolina: Pentland Press, Inc., 2000.

Wilson, John P. *Islands in the Desert. A History of the Uplands of Southeastern Arizona.* Albuquerque; University of New Mexico Press, 1995.

Articles

Anonymous, "The Mines of Freiberg in the Olden Times," *Scientific American*, Vol. IV, No. 20, May 18, 1861, pp. 316-17.

———, "Mining Schools of Germany; The Academy at Freiberg — Its History, Present Condition and Peculiar Advantages," *The Scientific American*, Vol. IV, No. 21, May 25, 1861, pp. 326.

Adams, W. H., "Copper Smelting in 1865," *Engineering and Mining Journal*, December 4, 1886, pp. 400-01.

Austin, L. S., "Fifty Years of Smelting in the West," *Mining and Scientific Press*, May 21, 1910, pp. 754-57.

Austin, Lawrence W., "Silver Milling at Charleston, Arizona," *Engineering and Mining Journal*, January 27, 1883, pp. 44-46.

———, "A Mexican Cupellation-Hearth," *Engineering and Mining Journal*, February 21, 1885, p. 121.

Bartlett, P. L., "Notes on Copper Smelting," *Engineering and Mining Journal*, December 16, 1882, pp. 319-20.

Birkinbine, John, "American Blowing Machinery," *Cassier's Magazine*, Vol. XI, No. 2, December 1896, pp. 109-29.

Blake, W. P., "Something About the Freiberg Mines," *Mining and Scientific Press*, October 12, 1867, p. 226.

Brodie, W. M., "Metallurgy of Native-Silver Ores of Southwestern Chihuahua," *Engineering and Mining Journal*, Vol. 101, No. 7, January 12, 1916, pp. 297-298.

Brook, Harry Ellington, "The Great Silver King Mine," *Los Angeles Mining Review*, Vol. IX, No. 12, March 23, 1901, p. 17.

Callan, J. C., "Centrifugal Air Compressors," *Mining Science*, July 15, 1909, pp. 50-51.

Christensen, Charles C., "An Early Type of Water-Jacketed Smelting Furnace," *Engineering and Mining Journal*, Vol. 104, No. 19, November 10, 1917, pp. 836-837.

———, "Designing Small Copper Smelting Plants," *Engineering & Mining Journal*, Vol. 99, No. 5, January 30, 1915, pp. 225-228.

Church, John A., "Notes on a Metallurgical Journey in Europe," *Engineering and Mining Journal*, various issues between September 24, 1872, to July 29, 1873.

Church, John A. "Concentration and Smelting at Tombstone, Arizona," *Engineering and Mining Journal*, April 16, 23, 1887.

______. "The Tombstone, Arizona, Mining District," *Transactions, American Institute of Mining Engineers*, Vol. XXXIII (1902).

Collins, Henry F., "On Adobe and other Makeshift Furnaces," *Engineering & Mining Journal*, July 11, 1903, pp. 51-53.

Daggett, Ellsworth, "Economical Results of Smelting in Utah," *Transactions of the American Institute of Mining Engineers*, Vol. II, May 1873-February 1874, pp. 17-28.

Douglas, James, "The Copper Resources of the Unitd States," *Transactions, American Institute of Mining Engineers*, Vol. XIX, 1890-91.

———, "Crushing and Pulverizing Machines," *Cassier's Magazine*, Vol. XIII, No. 6, April 1898, pp. 496-518.

———, "Bessemerising of Copper and Copper Mattes," *Cassier's Magazine*, Vol. XXII, No. 3, July 1902, pp. 314-22.

Dowlen, E. J., "Early Lead Smelting in Western Camps," *Pacific Coast Miner*, February 14, 1903, pp. 120-21.

Earl, Phillip, "Nevada's Italian War," *Nevada Historical Society Quarterly*, Vol. 12, No. 2, Summer 1969, pp. 47-87.

Egleston, T., "The Manufacture of Charcoal in Kilns," *Transactions of the American Institute of Mining Engineers*, Vol. VIII, May 1879 to February 1880, pp. 373-397.

Gmehlino, A., "Smelting in Adobe Furnaces and Refining in Mexican Vaso," *Mining and Scientific Press*, January 3, 1885.

Hahn, O. H., Anton Eilers, and R. W. Raymond, "The Smelting of Argentiferous Lead Ores in Nevada, Utah, and Utah," *Transactions*, American Institute of Mining Engineers, Vol. I, 1871-72, pp. 91-131.

———, "Development of Silver Smelting in Mexico," *Transactions of the Institution of Mining and Metallurgy* (London), Vol. VIII, 1899-1900, pp. 231-299.

Hattori, Eugene M., and Marna A. Thompson, "Using Dendrochronolgy for Historical Reconstruction in the Cortez Mining District, North Central Nevada," *Historical Archaeology*, Vol. 21, No. 2, 1987, pp. 60-73.

Howard, F. T., "Manufacture of Refractory Materials for Furnaces," *Mining Science*, March 9, 1911, pp. 263-65.

Ingalls, W. R., "Early Mining in Utah," *Engineering and Mining Journal*, Vol. 98, No. 6, August 8, 1914, pp. 248-253.

Ives, Ronald, L., "The Cochran Coke Ovens," *Journal of Arizona History*, Vol. 13, No. 2, Summer 1972, pp. 73-81.

Jernegan, J. L., "Lead and Silver Smelting in Chicago," *Transactions*, American Institute of Mining Engineers, Vol. II, 1874, pp. 279-95.

Keyes, W. S., "Notes on Western Lead Smelting," *Annual Report 8*, California State Mining Bureau, 1888, 804-31.

Lang, Herbert, "The Copper Resources of California. III. The Copper Belt of the Sierra Foothills," *Engineering and Mining Journal*, May 27, 1899, pp. 619-620.

Larson, Gustive O., "Bulwark of the Kingdom: Utah's Iron and Steel

Industry," *Utah Historical Quarterly,* Vol. 31, No. 3, Summer 1963, pp. 249-261.

Mathisen, Jean A., "Piedmont and the Charcoal Kilns," *True West,* Vol. 45, No. 5, May 1998, pp. 24-27.

Morton, J. H., " "The Coal Mines of Central Utah," *Engineering and Mining Journal,* February 3, 1877.

Notarianni, Philip F., "The Frisco Charcoal Kilns," *Utah Historical Quarterly,* Vol. 50, No., 1, Winter 1982, pp. 40-46.

Pepper, William, "Lead Poisoning," *Scientific American,* Supplement No. 235, Vol. X, July 3, 1880, pp. 3745-46.

Pettibone, Hawley, "Gas-Powered Plants for Mining and Smelting," *Cassier's Magazine,* Vol. XXII, No. 1, May 1902, pp. 48-55.

Pulsifer, H. B., "The Metallurgy of Lead," *Salt Lake Mining Review,* August 15, 30, 1912.

Reid, John A., "Ore Deposits of Copperopolis, California," *Mining World,* Vol. 27, September-October 1907, pp. 471-72; 513-15, 599-600.

Rickard, Forrest R., "History of Smelting in Arizona," *History of Mining in Arizona.* Tucson: Mining Club of the Southwest Foundation, 1987.

Robie, Edward H. "Metallurgical Plants in the Vicinity of New York," *Engineering and Mining Journal,* Vol. 109, No. No. 24, June 12, 1920, pp. 1299-1301.

Rolando, Victor R., "Loading Charcoal at Danby Station," *Technology and Culture,* Vol. 35, No. 2, April 1994, pp. 383-388.

Ross, Joseph V. H., "Charcoal War," *True West,* Vol. 40, No. 7, July 1993, pp. 37-41.

Sexton, A. H., "Lead Smelting in Scotland," *Engineering and Mining Journal,* February 23, 1895, pp. 175-76.

Shepperson, Wilbur, "Immigrant Themes in Nevada Newspapers," *Nevada Historical Society Quarterly,* Vol. XII, No. 2, Summer 1969, pp. 5-46.

Zeier, Charles D., "Historic Charcoal Production Near Eureka, Nevada: An Archaeological Perspective," *Historical Archaeology,* Vol. 21, No. 1, 1987, pp. 81-101.

INDEX